THE INTERPET
QUESTIONS & ANSWERS
MANUAL

THE
WATER
GARDENING
HANDBOOK

JOHN DAWES

INTERPET ![logo] PUBLISHING

Contents

AN ANDROMEDA BOOK

Copyright © 1998
Andromeda Oxford Limited

Reprinted 1999, 2002

Planned and produced by
Andromeda Oxford Limited,
11–13 The Vineyard, Abingdon,
Oxfordshire OX14 3PX,
England

Distributed to the pet and aquatic trades by
Interpet Ltd., Vincent Lane, Dorking,
Surrey RH4 3YX

Editors	Peter Lewis, Lauren Bourque
Art Editors and Designers	Martin Anderson, Chris Munday
Picture Research Manager	Claire Turner
Proofreader	Lynne Wycherley
Indexer	Ann Barrett
Production Director	Clive Sparling
Publishing Director	Graham Bateman

Origination by Acumen Overseas Pte. Ltd.,
Singapore

Printed by Polygraf Print, Prešov, Slovakia

ISBN 1–84286–072–0

Introduction

When I first became involved in outdoor aquatics in the mid-1950s, most enthusiasts referred to themselves as "pond keepers". Today, many of us regard ourselves as "water gardeners". What's in a name, and what has changed in the interim?

The difference is not just a question of semantics. In pond keeping, the emphasis was – and still is – very much on fish and water quality maintenance. Most of us did incorporate some plants in our early schemes, but they remained peripheral to the main attraction, the fish. Moreover, the selection of plants was extremely limited, typically including submerged species such as Pondweeds and Hornworts and marginals like Reedmace and Marsh Marigolds, while surface and floating plants were represented by the token Water Lily or Water Soldier. By contrast, the broader concept of water gardening considers plants as integral, essential components of a water scheme. This notion has been promoted by the huge range of plants for all areas of the pond that have become available in aquatic stores and garden centres. Fish are still important, of course, but for a growing number of enthusiasts they are seen as part of a totality, and sometimes even play a supporting role to the plants.

During the 1980s, an important innovation helped water gardening to expand considerably: the development of the small water feature. These self-contained units – originally devised for people who wanted a pond but did not have the space – became an instant hit. Within a short time, they were taken up by existing pond owners, and gained widespread popularity. Every season now sees the launch of countless new designs, and small water features of every conceivable shape appear at major horticultural shows.

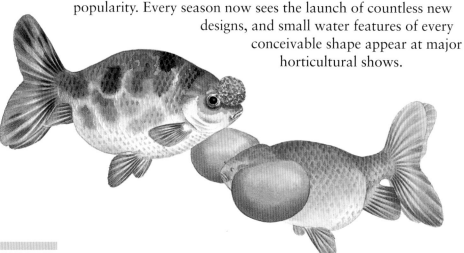

A selection of Goldfish, the most popular fish in the pond keeping hobby.

Water gardening is a leisure activity that can accommodate all tastes, and the possible permutations of ponds, bog gardens and small water features are endless. This book will provide you with essential information on constructing and maintaining a pond (whichever type you choose) and, I hope, will stimulate your imagination when you set about designing and creating your ideal set-up. Its broad coverage places equal emphasis on fish husbandry and plant care; in addition, a separate section deals with the native fauna that may visit the wildlife schemes that many pond owners the world over are now favouring. In the distinctive "question and answer" features, I have drawn on my many years' experience troubleshooting enthusiasts' problems. Finally, a select bibliography gives interested readers the opportunity to research individual topics in greater depth. Whatever your personal preference or level of expertise, there is something here for you.

A brief note on measurements; dimensions are given in both inches and centimetres, while liquid measurements are expressed in imperial, metric, and US units.

Thank you for buying this book. Whether you consider yourself a pond keeper or water gardener, I wish you every success!

JOHN DAWES

Pond Set-Up

THE CENTREPIECE OF EVERY WATER GARDEN – AND
the feature that distinguishes it from any other kind
of garden – is the pond. Many lifelong water gardeners
have first been attracted into this endlessly fascinating
hobby by the idea of owning a pond.

The size, shape and style of pond will vary according to a
number of factors, such as personal preference, suitability of site,
choice of flora and fauna, and the amount of space and money that
can be devoted to the enterprise. The first section of this book covers
the main types of ponds and their characteristics. Included is a
treatment of small water features, which have been growing steadily
in popularity in recent years. In addition, the practicalities of siting,
designing and constructing a pond are outlined, including the most
commonly used building materials and equipment. The aim is to
provide a broad base of knowledge, which can then be applied and
adapted to individual circumstances.

No two ponds are ever exactly the same. Even two identical,
prefabricated (moulded) ponds – bought on the same day and
installed at the same time alongside each other – will differ from
the very moment they are set up. This arises from the fact that it
is impossible to stock them with identical fish, plants or other
organisms, or to maintain them identically in every respect,
from the amount of food given to the turnover rate of
their filters. This infinite variety is just one of the things
that make water gardening such an enthralling and
rewarding activity.

▶ *The fruits of good forward planning, sound
construction and sensitive planting are evident in
this impressively varied pond.*

General or Mixed Ponds

Most new pond owners opt for a system that will accommodate more than one type of fish and also a selection of plants. This results in what is called a "general" or "mixed" pond. The attractions of mixed collections are many, and can be varied over time in such a way that even a lifetime of water gardening will not exhaust the myriad permutations on the basic theme.

The selection of animals and plants is the deciding factor in defining a mixed or general pond: as long as these are not exclusively of one type (e.g. Goldfish and water lilies), the system can legitimately be described as mixed.

There are no fixed rules about keeping a range of plants or fish in a single pond, nor about the style or size of the pond itself. Thus it is equally valid to house a mixed collection in a formal, circular cement pond (complete with central statue of a goddess), in an informal scheme incorporating a pond liner, or in a system that has been constructed around a prefabricated (moulded) pond design.

▶ *Many people choose to install a fountain in their pond. In many cases, the water jet is the principal attraction; here, however, mosaic ceramic pieces have been used to make a feature of the fountain itself.*

Design Considerations

There are a few basic requirements for the design of a mixed pond. How these are met and which materials are used are, of course, open to individual creative interpretation. It is this crucial and infinitely variable factor that makes this type of pond so exciting to design, install and maintain.

The latitude that mixed ponds allow must not, however, be taken to imply that "anything goes". Some rules apply just as strictly to mixed systems as to specialized ones. Above all, mixed ponds must cater properly for the needs of the plants and animals they contain, irrespective of shape, size or range.

Unless they are very large, few single designs can cater equally well for all possibilities. However, as most organisms have the inbuilt capacity to adapt, a general or mixed pond can house a wide range as long as a few essential elements are included.

◀ *A water garden with lush mixed vegetation. This is an extensive scheme, but with good forward planning, even a garden with limited space can accommodate a general pond.*

Structural Elements

In order to provide a suitable growing area for bog and shallow-water marginal plants, as well as places where the fish can bask in the sun and spawn, at least one shelf should be incorporated at a water depth of around 8–10in (20–25cm). By placing baskets or perforated containers either directly on this shelf or on a raised support, such as a brick, the water gardener can create a range of planting conditions, from moist soil for bog plants to several inches' depth of water for shallow-water marginals (see Pond Plants, pages 130–143).

Considering a General or Mixed Pond

Advantages
• Mixed systems allow a variety of fish and plants to be kept together.
• The huge range of designs means that mixed ponds can fit almost all gardens and patios.
• A mixed pond that is both attractive and provides adequate conditions for the pond inhabitants need not be expensive.
• Mixed pond systems can normally be installed in a relatively short time.
• It is not usually vital to have elaborate or expensive pieces of equipment in mixed ponds.

Disadvantages
• By definition, mixed collections prevent specialization, for example into keeping Koi or Goldfish.
• Due to the varied nature of general collections, you should take great care not to choose incompatible fish.
• The extensive range of fish and plants that are on offer for general or mixed ponds can easily lead to a temptation to overstock beyond the pond's capacity. This will harm your livestock's health.

▲ *The Japanese Water Iris* (Iris laevigata) *is an extremely popular choice of shallow-water marginal pond plant. The variety shown here is* 'Atropurpurea'.

Ease of access and exit for amphibians must be provided in the form of at least one gentle slope into the water. This can be constructed in various ways: a "beach" made of pebbles lying on a broad shelf; a length of suitable material such as coir (coconut matting) or pebble-dashed pond liner draped over the edge; or pondside plants overhanging or extending into the water.

Deep Areas in Mixed Ponds

Moving in from the edge of general or mixed schemes, deeper shelves will make suitable sites for cultivating deep-water marginals and surface plants, such as Water Hawthorn (*Aponogeton distachyos*) or water lilies. In the very centre of the pond, there should be an area where the water is at least 18in (45cm) deep – and preferably deeper.

Q & A

● *What are the ideal surface dimensions for a mixed pond?*

... It is difficult to stipulate "ideal" dimensions for a pond; a large number of variable factors – such as the space and funds available, and the type and number of fish and plants chosen – all play a role in determining its ultimate size. In general, because larger ponds are less likely to experience rapid fluctuations in water quality and temperature, these are usually to be recommended. In any event, the absolute minimum surface area for any mixed pond should be around 30–35sq ft (2.8–3.25sq m), or preferably 50sq ft (4sq m). This area can be distributed over any shape that is required or desired – see Pond Design, pages 30–31. If the pond is any smaller than this, it may prove difficult to maintain in peak condition and can require constant, labour-intensive attention, especially if it is fully stocked with medium-sized or larger fish. On the other hand, if you are only intending to keep small fish, these will not cause water quality fluctuations to the same extent as larger specimens, and so are much easier to manage.

Some Fish and Plants for the Mixed Pond

FISH
Goldfish (*Carassius auratus*)
Crucian Carp (*Carassius carassius*)
Prussian Carp/Gibel (*Carassius gibelius*)
Koi (*Cyprinus carpio*)
Golden/Blue/Marbled Orfe (*Leuciscus idus*)
Golden Rudd (*Scardinius erythrophthalmus*)
Roach (*Rutilus rutilus*)
Green/Golden/Red/Red and White Tench (*Tinca tinca*)
*Golden Fathead Minnow (*Pimephales promelas*)
European Minnow (*Phoxinus phoxinus*)
*Southern Redbelly Dace (*Phoxinus erythrogaster*)
*Red Shiner (*Cyprinella lutrensis*)
Dace (*Leuciscus leuciscus*)
Gudgeon (*Gobio gobio*)
*Golden Medaka (*Oryzias* spp.)

PLANTS
Astilbe (*Astilbe* spp. and vars.)
Bugbane (*Cimicifuga simplex*)
Giant Rhubarb (*Gunnera manicata*)
Hosta (*Hosta* spp. and vars.)
Monkey Flower/Musk (*Mimulus* spp. and vars.)
Globe Flower (*Trollius* spp. and vars.)
Primulas (*Primula* spp. and vars.)
Water Iris (*Iris laevigata; I. versicolor*)
Houttuynia (*Houttuynia cordata*)
Sweet Flag (*Acorus camalus*)
Water Plantain (*Alisma* spp.)
Cotton Grass (*Eriophorum angustifolium*)
Water Forget-Me-Not (*Myosotis scorpioides*)
Dwarf Reedmace (*Typha minima*)
Brooklime (*Veronica beccabunga*)
Marsh Marigold (*Caltha palustris*)
Papyrus (*Cyperus papyrus*)

Water Purslane (*Ludwigia palustris*)
Arum Lily (*Zantedeschia aethiopica*)
*Lotus (*Nelumbo* spp. and vars.)
Water Lily (*Nymphaea* spp. and vars.)
Water Hawthorn (*Aponogeton distachyos*)
Fairy Moss (*Azolla* spp.)
*Water Hyacinth (*Eichhornia crassipes*)
*Frogbits (*Hydrocharis morsus-ranae* and *Limnobium laevigatum*)
Duckweeds (*Lemna* spp.)
*Water Lettuce (*Pistia stratiotes*)
Water Soldier (*Stratiotes aloides*)
Pondweeds (*Egeria, Elodea, Lagarosiphon*)
Curled Pondweed (*Potamogeton crispus*)
Milfoils (*Myriophyllum* spp.)

*These fish and plants cannot be considered frost-tolerant or temperate-winter-hardy and are better suited, either for tropical or subtropical ponds, or only as temporary (summer) residents in more temperate latitudes.

◀ *Golden Orfe* (*Leuciscus idus*)

● *To what depth should a general or mixed pond extend?*

Ponds in temperate areas should have a water depth of at least 18in (45cm) to provide adequate winter protection for the fish. An equally important factor, but one that is often overlooked, is that ponds in areas with high summer temperatures should also be at least this deep to minimize the risk of overheating. A depth of around 18in should be sufficient for most small and medium-sized fish, except in extremely cold areas. However, if large fish are to be kept, the depth should be increased accordingly. Ponds containing large carp or Koi, or fish of an equivalent size, should be at least twice this depth – 36in (90cm).

This deep area is absolutely indispensable, for a variety of reasons. Firstly, it forms a place of refuge for fish, both large and small, whenever they feel threatened. It will also create a region of comparatively warm water during freezing weather conditions, which will allow the fish to hibernate in relative safety. On the other hand, in hot climates it is likely to be the coolest (and deepest) area, where temperate fish species will feel most comfortable. The deepest part of the pond is also suitable for larger water lilies.

Mixed Pond Inhabitants

By the very nature of the general or mixed pond, the inhabitants are extremely varied. They may include flora and fauna as diverse as bog, deep-water and shallow-water marginal plants, amphibians, fish of various species and varieties, and water lilies, together with oxygenating and floating plants. One thing that you must ensure is that fish species are biologically compatible, so as to avoid predation.

Koi Pools

THE IMAGE THAT MOST PEOPLE HAVE of Koi pools is of a formal arrangement consisting of a cement, brick or block pond with crystal-clear water, no plants and an extensive selection of brilliantly coloured, large Koi. Indeed, systems that are set up mainly (or exclusively) for keeping and displaying Koi still tend to conform to this traditional image of Koi keeping. These pools also frequently incorporate decorative objects, such as statues and lanterns, as well as bonsai trees, cascades and similar features.

However, as the interest in gardening in general – and water gardening in particular – has continued to expand, there has been a tendency for Koi and garden enthusiasts to combine their hobbies, often to spectacular effect. It is therefore now more common for those who are both keen gardeners and Koi owners to design their entire garden around a Japanese theme to harmonize with their Koi pond. As this trend has grown, so has the availability of lined and prefabricated (moulded) Koi ponds; while cement, brick and block designs still dominate the market, other types of pond are no longer as rare as they once were.

Koi ponds are far more exacting in their requirements than general or mixed ponds, especially as regards their filtration and drainage systems. Their design and construction demand a level of expertise that exceeds the capabilities of most first-time pond builders. Before proceeding with the construction of a Koi pool, therefore, water gardeners would be well advised to consult specialist Koi literature (see Further Reading, page 202) and recognized experts in the field. Membership of a Koi society is also strongly recommended.

Basic Requirements

The wide range of materials that can be used for Koi pool construction means that such systems can be built to suit personal preferences and site constrictions, while at the same time catering for the needs of the fish. Among these requirements, swimming space (in terms of overall surface size of the pond) and water depth are important factors that need to be considered from the outset.

Bearing in mind that Koi will grow into large fish that can comfortably exceed 24in (60cm) in length, with corresponding body depth, weight

▼ *Keeping Koi need not entail creating a barren pond completely devoid of plants. This attractive setting for a Koi pool incorporates general garden plants around the pond edge and an ornamental stone pagoda. Yet, as is fitting, the focus of the pond remains the fish themselves.*

Considering a Koi Pool

Advantages
- A properly constructed, stocked and managed Koi pool is a very beautiful feature.
- Koi, and in particular large Koi, can become quite tame and thus develop into genuine "companion" pets.
- Owing to their dimensions, Koi pools can accommodate large numbers of fish.
- Owning a Koi pool encourages the fish-keeper to acquire an in-depth knowledge of fish in general, and carp and Koi in particular.
- Koi pool owners often form lasting friend-ships with each other, and develop a strong sense of being part of a very special branch of the fishkeeping community.

Disadvantages
- Koi pools can be expensive to construct, install and maintain.
- They can take up an inordinately large amount of space.
- They require more regular attention than mixed or general ponds.
- Repairs can be awkward or expensive to undertake.
- In certain areas and countries, Koi pools have attracted the attention of "rustlers" (i.e. fish thieves), particularly if the collection includes valuable pedigree specimens.

● *How deep should a Koi pool be?*

Koi pools should be at least 4ft (1.2m) deep, but serious Koi keepers always aim for a greater depth – around 5ft (1.5m) – as their minimum. The optimal depth (bearing in mind water clarity) is often quoted as c.8ft (2.4m), yet this need not be uniform throughout the pool.

● *What is the minimum surface area for a Koi pool?*

As with mixed ponds, it is hard to give a definitive minimum size for Koi pools, since this depends on variable factors such as fish numbers and sizes, type and efficiency of water-management systems, and feeding frequency. Yet around 95–110sq ft (9–10sq m) is the very smallest acceptable surface area. Even a pond this size will only be able to house a few large specimens, which will not make for a very impressive display. A minimum surface area of 150–160sq ft (14–15sq m) would thus be preferable.

and metabolic demands, small pools and Koi are, by definition, mutually exclusive (see Q&A opposite, and box on Suggested Koi Pool Conditions, page 14).

While general or mixed ponds incorporate one or more shelves on which to set plant baskets, pools that are constructed primarily to house a Koi collection are usually designed without shelves, or with one wide and deep shelf, as well as vertical sides (not sloping). Another difference between Koi and mixed ponds is that with the latter, the deepest point is usually in the centre,

- Koi pools should be as large as possible. No strict rules can be given regarding maximum surface dimensions, since much depends on stocking levels and water-management techniques. However, as a rough guide, surface dimensions should never be less than 9x9ft (3x3m), and 4ft (1.2m) in depth.

- The filtration system should be the most efficient that is affordable. If the system required to keep toxins like ammonia and nitrites under control is beyond the budget, the number of fish should be reduced.

- The filter system should contain at least a settlement section, a mechanical component and a biological one. Additionally, an ultraviolet (UV) sterilizer/clarifier should be seriously considered.

- Drains are desirable, but not essential. If drains are not installed, an alternative method of keeping the water free from solid wastes, e.g. a pond vacuum, should be obtained.

- Aeration levels should be kept high, either by using waterfalls or cascades, or by means of air pumps and aeration devices such as venturis (simple fixtures that create a turbulent, oxygen-rich mix of air and water).

- A vortex chamber should be included to intercept the solid waste and debris that Koi produce in copious amounts. These are large, cylindrical containers with a cone-shaped base, an inlet pipe at the top and an outlet at the bottom. The inlet pipe enters the chamber parallel to the internal wall of the cylinder, and discharges the pond water so that it races around this inside wall, gradually slowing down as it is dragged towards the centre of the cylinder. As this happens, the solid wastes introduced with the pond water settle at the bottom of the vortex created by the swirling water column, from where they can be collected and disposed of. The solid-free water is then channelled into the pond filter, where it undergoes further treatment. Vortex chambers are particularly popular among Koi keepers, but their use is also growing in more general areas of pond keeping.

▶ *A vortex chamber. Units like this substantially reduce the pollution load on later filtration stages.*

while the deepest point in a Koi pool is often at one end, as in a swimming pool. This deep area can be quite large. One common feature of Koi pools that is rare in other types of pond is the presence of one or more drains (depending on the size of the pool). The main reason for including these is that Koi produce large quantities of solid wastes, which can be difficult to remove from deep water, other than by "flushing" them out of the system.

Koi Pool Filtration

The filtration needs of most Koi pools are serviced by purpose-built external units incorporating multiple filter beds. Other types of filter are often used to supplement these. They are basically of two types, neither of which can be considered as a complete system in itself, but which both perform very useful supporting roles:

Sand filters – these robust spherical filters are, as their name implies, packed with sand grains. Water is pumped through under pressure and emerges with added clarity before returning to the pool.

Cartridge filters – these tall cylinders contain a long cartridge of a cellulose-based mechanical

medium or activated charcoal. Water is pumped through the cylindrical chamber until the medium clogs up and is either cleaned or replaced.

A problem that can arise in sparse, Koi-only pools is that the filtration system will produce an excess of nitrates, since there are no plants to take up these nutrients. As a result, blanket weed or the nuisance algal bloom known as "green water" will profit from this superabundance of plant food, and can overrun the pond. One way in which specialist Koi keepers overcome this problem is to operate a so-called "plant filter". This entails passing the pond water, after normal filtration, through an additional filter bed separate from the main pond, where oxygenating and other plants take up excess nitrates before the water is returned to the pond. UV clarification will also help.

▶ *Clean water is essential for the health of the fish and for fish keepers' enjoyment of their livestock. Here, two Ghost Koi are seen feeding on floating flake food.*

▼ *A multi-chambered Koi pool filter. These installations are gravity-fed, and comprise a series of progressively finer filter media.*

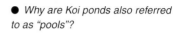

● *Why are Koi ponds also referred to as "pools"?*

... The traditional plantless set-ups that were built to accommodate Koi tended to resemble swimming pools rather than garden ponds. Consequently, the term "pool" came to be regarded as a more apposite description of them. Nowadays, however, Koi are kept in such a wide range of settings that depart from the more clinical "pool" arrangement that the word "pond" has reasserted itself. So, "pool" and "pond" are now virtually interchangeable, and equally acceptable, terms.

▼ *In the absence of oxygenating plants, oxygen levels in Koi pools can get very low. As a result, the fish will suffer and the aerobic filter bacteria will not multiply efficiently. To overcome this problem, Koi keepers use air pumps.*

Wildlife Ponds

IN THEORY, A WILDLIFE POND IS ONE BUILT entirely from natural materials and containing only animals and plants native to the area. In practice, though, the vast majority of such ponds are compromises, which fall short of this narrow definition, but which nevertheless are far enough removed from mixed ponds and Koi pools to qualify as wild.

Growing interest in wildlife pond keeping has led many people to encourage local flora and fauna to establish themselves in their gardens – frequently alongside exotic or cultivated species. One important consequence of this is that many of the more informal water garden schemes now act as refuges for some amphibian and other species whose survival in the wild is threatened.

Construction Materials

Ponds were once considered wild only if they were formed in a natural, water-holding hollow, or in an artificial excavation lined with a natural material, such as clay. Nowadays, many different kinds of lining materials are regarded as valid for wildlife ponds, such as liners, cement, bricks and blocks. Even prefabricated (moulded) ponds can form the basis of a wildlife scheme.

▶ *A wildlife pond in a secluded location. This scheme incorporates a profusion of wild-type plant species, and demonstrates well the allure of an informal water garden arrangement.*

Clay-lined structures in artificially excavated depressions are the most natural form of pond that the majority of gardeners can hope to achieve. Such ponds have undergone a revival in recent years. The traditional construction method is from clay that is repeatedly trodden down until it forms a water-retaining layer. This long and exhausting process, which is referred to as "puddling", demands considerable knowledge, stamina and patience. Unsurprisingly, therefore, many so-called "clay" ponds are built from overlapping layers of self-sealing, bentonite-impregnated liner. In fact, this results in a more effective and durable waterproof lining than can be obtained with genuine puddled clay.

Wildlife Pond Inhabitants

As with construction methods and materials, the approach that is adopted by most water gardeners and pond keepers to stocking their wildlife schemes is flexible enough to include wild-type

Considering a Wildlife Pond

Advantages

- In many cases, wildlife ponds represent the only safe havens for local aquatic wildlife.
- More than any other type of water scheme, wildlife ponds allow us to get close to natural aquatic fauna and flora throughout the year
- Wildlife ponds are relatively inexpensive to install and stock.
- Once established, such ponds attract to the garden wildlife other than aquatic or amphibious species (e.g. birds, hedgehogs).
- As the amphibian population grows, spawn can be passed on to other wildlife pond enthusiasts, or used to restock natural ponds locally.

Disadvantages

- As wildlife ponds are not usually serviced by pumps and filters, the water can become cloudy and laden with sediment.
- Accumulation of leaves and debris necessitates a complete clean-out every 2–3 seasons.
- Owing to their saucer-shaped profile, wildlife ponds with an adequate water depth in the centre take up a lot of garden space.
- Fish stocking levels have to be lower than in a general pond.
- Plants not grown in containers can be difficult to keep under control, especially specimens in the deep, central area of the pond.

animals and plants, regardless of whether they are strictly "native" to the region where the pond is located. Thus, a wildlife pond in the USA might reasonably contain wild-type Goldfish (*Carassius auratus*) – which are indigenous only to China and parts of Siberia – while inhabitants of a European pond could include North American Red Shiners (*Cyprinella lutrensis*) and wild-type Japanese Medakas (*Oryzias latipes*). The only animals and plants to be excluded are those that have been developed commercially or artificially into forms not found in the wild. So, an appropriate choice of water lily for a wildlife scheme would be the Yellow Pond Lily (*Nuphar lutea*) rather than a cultivated variety, such as *Nymphaea* 'Gloire du Temple-sur-Lot'. Yet even on this latter point, some latitude exists in practice, so long as there is no major departure from the overall concept of a wildlife pond. Ultimately, the main criterion is that the balance of the pond should be kept on the "wild" side.

Shape, Size and Distinctive Features

In their general shape, wildlife ponds differ from other types in having gently sloping sides all round, no shelves and an area in the centre that should be at least 18in (45cm) deep, and preferably 24in (60cm). This results in a saucer-shaped profile that provides a smooth transition from dry land, through a boggy/marshy area, to a progressively deeper, aquatic zone that can accommodate the varying needs of the range of plants and animals with which the pond is stocked.

Wildlife ponds that conform to this ideal shape are necessarily large. Even a relatively steep slope of 30° will give a diameter of 13ft (4m) or more, depending on the size of the central deep area. The bare minimum surface area is 30–35sq ft (2.8–3.25sq m); but note that even this will not reach the recommended minimum water depth.

One feature that all wildlife ponds should possess, irrespective of their construction materials, is a heavily planted shore zone. Among the many functions this performs, the most important are that it provides: an easy (and relatively safe) entry and exit point for the many creatures that use the pond on a temporary (though not necessarily short-term) basis; perches for airborne

FEATURES OF A WILDLIFE POND

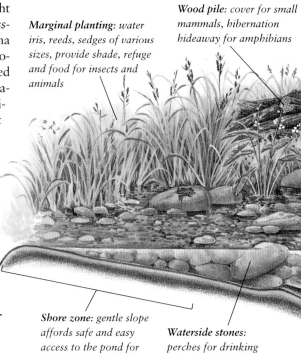

Marginal planting: water iris, reeds, sedges of various sizes, provide shade, refuge and food for insects and animals

Wood pile: cover for small mammals, hibernation hideaway for amphibians

Shore zone: gentle slope affords safe and easy access to the pond for wildlife

Waterside stones: perches for drinking birds, basking spots for amphibians

visitors such as dragonflies and damselflies; and sites onto which the fully-grown aquatic nymphs of these insects can climb and wait until the adult winged stage is ready to emerge.

Adapting Prefabricated Ponds

Saucer shapes are relatively easy to create using either pond liners, puddled clay or concrete, but not with prefabricated (moulded) ponds, since these come with built-in shelves and steeply sloping sides. This does not completely exclude moulded designs from use as wildlife ponds, but they should be adapted to suit wild creatures that find steep-sided ponds difficult to manage. This can be achieved in many ways, such as creating pebble beaches on the shallowest shelves, draping fibrous material (e.g. coir or coconut matting) into the water, arranging planting baskets with bog plants or shallow-water marginals along the top shelves and as close to the pond edge as possible, or fitting special wildlife entrance and exit ramps.

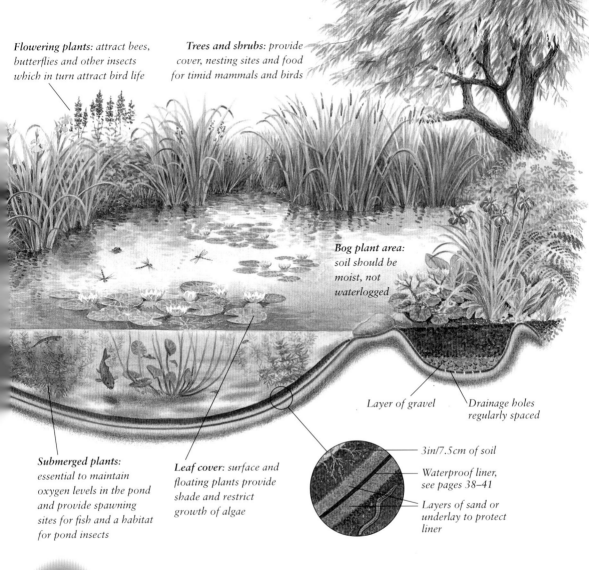

Flowering plants: *attract bees, butterflies and other insects which in turn attract bird life*

Trees and shrubs: *provide cover, nesting sites and food for timid mammals and birds*

Bog plant area: *soil should be moist, not waterlogged*

Layer of gravel

Drainage holes regularly spaced

Submerged plants: *essential to maintain oxygen levels in the pond and provide spawning sites for fish and a habitat for pond insects*

Leaf cover: *surface and floating plants provide shade and restrict growth of algae*

3in/7.5cm of soil

Waterproof liner, see pages 38–41

Layers of sand or underlay to protect liner

Q&A

● *Do the still-water conditions in wildlife ponds cause any problems?*

... In still conditions, heat absorbed at the surface of the water does not easily disperse through the deeper layers. As a result, stratification (or "layering") occurs. This can pose a problem for some of the more tropical, less hardy types of fish kept in temperate zones. Along with temperate species, these fish are attracted to warm water in the shallows during the spring and summer months. Yet if they are suddenly forced to pass from this zone into a deeper, colder region without any metabolic adjustment – for example when fleeing from a predator – they can suffer abrupt chilling. If these fluctuations in body temperature keep recurring, they can result in several problems, such as swimbladder malfunction and whitespot – see Health Management, page 77.

● *How can I create a bog garden?*

The essential thing to aim for in a bog garden is moist soil that is neither waterlogged nor runs any risk of becoming oxygen-deficient (anaerobic), which would cause toxic, polluted conditions. One of the two ways in which this can be achieved is shown in the accompanying diagram. Here, an extension of the pond liner is used to create the bog area, with drainage provided via a bottom layer of gravel and regularly spaced holes in the liner itself. Alternatively, the bog garden can be constructed separately, i.e. isolated from the pond water supply. In such cases, a separate source of water needs to be provided, for example by means of a dripping pipe laid (but concealed) on or below the surface, or actually embedded within the bottom gravel layer. Drainage holes must also, of course, be provided.

Small Water Features

THE SMALL WATER FEATURE WAS ONCE AS RARE as it is now popular. These eye-catching additions to the garden scheme come in a huge variety of different forms, and can be bought as manufactured items, or fabricated at home; the only limiting factor is your imagination!

A small water feature can be anything from an old kitchen sink or a wooden half-barrel adapted to take small fish, plants and assorted accessories, to a more substantial millstone, pebble or boulder fountain.

The main components of millstone fountains are large, circular stones with a hole in the centre. Millstones were originally used in windmills and watermills for grinding wheat, rye and other grains into flour. (If your local aquatic store cannot supply one, they may be able to advise you of alternative sources, such as salvage companies.) In the water feature, the millstone is positioned above a reservoir, which is equipped with a pump whose spout projects through the hole in the millstone. Pebble fountains have a rust-resistant mesh overlaid with a layer of pebbles through which the pump outlet extends, and onto which the spray splashes on its return to the reservoir. Boulder fountains work on the same principle.

Growth in Popularity

In the early to mid-1980s, millstone and pebble fountains were the only types of small water feature available, and even these were not commonly seen in people's gardens. The market for small water features expanded rapidly once the basic concept of a reservoir, a pump and some form of upper covering or decorative surface had been adapted for individual use and offered in a range of attractive styles.

Initially, small water features were the ideal solution for people living in apartments with only a small balcony or terrace, or house owners with insufficient space to install a pond. However, they swiftly excited the interest of traditional pond owners and water gardeners as well, even those with large systems. Today, as a result of consumer demand for something out of the ordinary to grace their gardens, designers and manufacturers are constantly coming up with fresh ideas and creations.

Whether the water feature is going to comprise your entire water garden, or whether you wish to use it as part of a larger scheme, there is sure to be one that will suit your needs. Keep in mind the advantages and disadvantages listed below.

Considering a Small Water Feature

Advantages

• Small water features can bring outdoor aquatics within the reach of every hobbyist, particularly those who do not possess a suitable site for a pond.

• The range of possible designs is infinite.

• Stunning displays can be created in the smallest of spaces.

• The availability of self-contained kits on the mass market results in "almost instant" water gardening.

• If no appropriate kits are available, the necessary components for a water feature are easy to obtain, devise or construct.

Disadvantages

• The small size means that the range of fish and/or plants is limited. Many designs are not suitable for either fish or plants; those chosen must be in keeping with the space available.

• Alternative accommodation for the fish must be provided in areas that experience severe winter conditions.

• In hot countries, shade is needed during the late morning and early afternoon.

• Regular checks must be carried out to ensure that the submersible pump is never exposed to the air. These systems lose a lot of water by evaporation and splashing.

▲ *This whimsical water feature is set in a wooden half-barrel. There is a growing demand among water gardeners for such unusual designs.*

Q&A

● *Are any containers unsuitable for use as small water features?*

... Any container that can release toxic compounds – such as wooden barrels that have stored preservative chemicals – is totally unsuitable for fish and/or plants. Containers that can corrode, such as iron, are unsuitable on a long-term basis, even if there are no fish or plants. Also avoid containers that cannot be made waterproof.

● *What is the size limit for a small water feature?*

Most are less than 5ft (1.5m) across, and many are smaller, especially free-standing models. Pebble, millstone and boulder fountains usually feature at the top end of the size range, and the overall dimensions of a combined display can be much bigger. At the other end of the scale, there are table-top or shelf water features, which are designed for indoor use and may have a base measuring just 12in (30cm) across.

▲ *The "deer scarer" (Shishi odoshi) is a traditional Japanese bamboo and stone feature that is designed to make a noise as the central tube tilts under the weight of water flowing in it. It has become popular with pond keepers as a purely aesthetic object.*

Building a Small Water Feature

Small water features may be either home-built or bought as a kit from a manufacturer. They can take virtually any form you like; the only factors that they have in common are their relatively small size and the presence of water. If you are using a kit, everything is provided and the whole system can be set up in a very short time – sometimes even in minutes. If you are building your own feature from individual components, there are four basic elements, listed below.

With these elements, a vast range of different designs can be created, either raised or sunken. Keen Do-it-Yourself (home improvement) enthusiasts can take designs a stage further by constructing the reservoir itself using cement, or by lining an excavation of a chosen shape or size with a pond liner and then adding the other "ingredients". The four essential elements are:

A large container to act as the reservoir, which must be waterproof. If it is not, then you should surround it with a piece of pond liner or apply a proprietary sealing compound.

A small, submersible pond pump capable of raising an unobstructed jet of water through the outlet pipe to achieve a height of about 6–12in (15–30cm; with fountain attachment removed). This is adequate for most designs of small water feature, which rely on an unmodified, solid vertical stream of water that falls back on itself. If you want to create other types of display, the size of the pump must be matched accordingly.

▲ *An old water pump, a milk pail and a half-barrel have been combined to good effect in this rustic running water feature.*

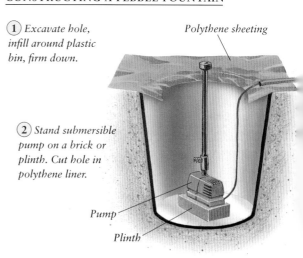

You will need...

- Large plastic container to act as a reservoir
- A small submersible pump with electrical wiring and waterproof connectors
- Coated wire mesh
- A length of delivery pipe and fountain
- Heavy duty polythene or small piece of pond liner
- A brick or plinth to stand the pump on
- Equipment: spade and other digging tools, wire cutters
- Pebbles, selected for their visual appeal when wet

CONSTRUCTING A PEBBLE FOUNTAIN

① *Excavate hole, infill around plastic bin, firm down.*

Polythene sheeting

② *Stand submersible pump on a brick or plinth. Cut hole in polythene liner.*

Pump

Plinth

Space around the pump to allow it to work efficiently. If the feature is a pebble fountain or something similar that could potentially obstruct the free flow of water, the necessary space can be created in a number of different ways. Here are two possibilities:

(i) a rustproof mesh supported on bricks, which will create a false bottom, allowing clear space above the pump, but providing a suitable hole through which the outlet pipe can fit.

(ii) a large, upturned plastic bowl – for example, a kitchen washing-up bowl – with a hole cut out of the bottom to accommodate the pump outlet.

A selection of pebbles or small rocks to form the top layer of the display.

These practical tips are intended to serve as general guidelines, and all the various items can be adapted to suit individual designs. For example, in the particular set-up of a pebble fountain illustrated below, the non-corrosive mesh is not supported on a false bottom to create clear space around the pump outlet, but rests directly on top of the reservoir itself. Such an arrangement works fine, as long as the pebbles are placed carefully so as not to block the outlet. Similarly, the water display can be made as elaborate as the individual water gardener wishes, through the use of different fountain jets attached to the pump outlet. Medium-sized or large pebble fountains can look particularly impressive when a powerful pump is used to produce a tall, "solid" simple water column.

Q&A

● *What are the relative merits of sunken and raised small water features?*

In regions of the world where winter and summer temperatures are extreme, a sunken water feature will remain warmer for longer during cold weather, and cooler for longer during hot spells. This could be an important consideration in areas or circumstances where fish cannot easily be transferred to alternative accommodation for winter, or where exposure to heat cannot be avoided. Some types of small water feature have their reservoirs located below soil level, for example most millstone and pebble fountains, while others, such as the Japanese "deer scarers" (*Shishi odoshi*) and hand-washing troughs or basins (*Tsukubai*) are strictly above-ground designs. Personal taste plays a major role, and it might simply be the attractiveness of the receptacle itself that causes you to decide in favour of one or other of the options.

● *Do small water features require much maintenance?*

Most certainly. Even in the absence of any plants or fish, debris willl accumulate over time, and this will eventually clog up the foam (or other type of) pre-filter on the pump. Fountains may also trap debris, which will alter the spray pattern, while algae can form in the water and on the rocks or pebbles. When plants or fish are added to suitable small water features, the usual management tasks that apply to these when they are grown or kept in conventional ponds will still apply. Therefore, routine maintenance of small water features is as essential as it is for full-sized ponds, but on a lesser scale.

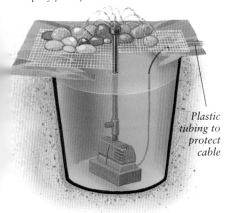

(3) *Overlap mesh by 4in/10cm on all sides. Arrange pebbles to best effect to catch spray from fountain.*

Plastic tubing to protect cable

▲ *A medium-sized or larger pebble fountain installed on a patio can accommodate a spray fountain.*

● *Can glazed containers, such as ceramic sinks, be coated to make them more attractive or suitable as small water features?*

Yes; old or discarded glazed sinks and other similar containers can be easily and effectively coated. The first step is to roughen or score the glazed surface (and a short distance down into the container) with an abrasive material, such as sandpaper used for metal surfaces. Once the surface is rough, it will take a layer of bonding agent which, while still sticky, is then coated with either a commercially available product, such as "hypertufa" (which simulates the porous, calcium-rich volcanic rock known as tufa) or a moist mixture of sand, cement and peat. The amounts of sand and peat can be varied to produce a coating of the desired colour and texture, but the overall mix should contain about 1 part cement to about 5 parts desired sand/peat blend. This mixture can be applied directly onto the tacky adhesive or, preferably, on top of a previous (and still wet) coat of a 1 part sand:1 part cement mix (wear protective gloves). Although it is unnecessary to coat the whole of the inside of the container, extend the coating at least 3in (7.5–8cm) down from the rim, so that the ceramic interior of the vessel will not be visible above the final waterline. Once the coating is completely dry, the sink is ready for use, although it won't be safe for fish until the toxic compounds contained within the cement have been allowed to leach out.

● *When converting a glazed sink into a small water feature, what is the best way of sealing the plug hole?*

Use the original plug, or buy a new one that fits snugly. Remove all the metal fittings (e.g. the chain and the fastener on the plug itself). Then spread a generous coat of silicone sealant around the hole and the edge of the plug. Press it firmly into position until the excess sealant oozes out; don't scrape this surplus off, but spread it over the top of the plug to form a good, watertight seal. Leave for a day before filling to allow the sealant to cure thoroughly.

▶ *This striking glazed planting pot has been turned into a small pond. The miniature water lily variety growing in it is* Nymphaea odorata 'Minor'.

Plants for a Container Pond

Surface Plants
Water lilies sold as miniature, pygmy or dwarf varieties.

Floating Plants
Small-leaved types, e.g. Duckweeds (*Lemna* spp.), *Azolla* spp., or *Salvinia* spp. Thin regularly to let some light penetrate to the bottom of the pond.

Marginal Plants
Small types, e.g. Water Forget-Me-Not, some Monkey Flowers, Bog Arum, Brass Buttons, Houttuynia, and Dwarf Reedmace.

Submerged Oxygenators
Most commonly available species, but curb their growth to stop them from choking the pond.

◄ *Even the most restricted corner of your garden or patio can be used as the site for a small water garden. Here, an old ceramic kitchen sink has been recycled to provide a home for Dwarf Reedmace and Irises.*

A Mini-Pond on Your Patio

Small still-water mini-ponds also make stunning displays, which can be every bit as beautiful as their moving-water counterparts. Such set-ups are particularly appropriate for patios or other sites where electricity cannot easily be provided or where, for whatever reason, a moving-water feature is unsuitable or undesirable. Bear in mind, however, that this type of feature will not give you the attractive sound of flowing water in your garden.

One advantage that still-water features have over moving-water types is that water loss from them through evaporation is far lower. On the other hand, the risk of stratification during hot weather – namely, the layering effect of warm water above cooler water, which creates a temperature gradient from the surface to the bottom of the pond – is considerably higher. This will not be a problem if the mini-pond is meant to accommodate only plants, but can be a serious drawback if fish are going to be introduced.

When buying fish for a still-water feature, choose only those species and varieties that are known to be hardy. Small specimens of the more basic varieties of Goldfish, such as the Common Goldfish or Shubunkin, are probably the wisest purchases. However, other small cyprinids or certain species of "Coldwater" Tropicals (see pages 108–115) may also be suitable, assuming that ambient weather conditions fall within their tolerance range.

Streams, Waterfalls and Cascades

VERY FEW GARDENS HAVE NATURAL STREAMS, waterfalls or cascades. Where such features do exist, they can be incorporated into the overall water garden design, but just how this is done depends on several factors. If the design entails redirecting the water flow, either upstream or downstream, from the section that crosses the property, then this must be discussed with the other parties affected. Local laws must also be consulted before any alterations are made.

Running water features can be purchased as ready-made, easy-to-assemble units, or may be home-built. Of the two options, the ready-made designs are far easier and quicker to install.

In either case, the ground should be prepared and the guidelines adhered to (see under the various pond types in the following chapters). They can be adapted as necessary, for example by allowing suitably large overlaps of liner, minimizing creases, removing sharp objects from the subsoil, and lining the subsoil.

Streams, waterfalls and cascades are almost invariably designed as part of an overall water garden scheme. When planning a running water feature, there are several important points to keep in mind.

Streams
The gradient created for a stream should be between 10° and 30°; any shallower than 10° may result in less flow or turbulence than is required, while more than 30° can result in either a torrent or extremely shallow water. These figures apply to an unrestricted stream that runs in a single stretch from its origin to the pond. Where the flow of water is interrupted on the way down, for example by the presence of several small pools (a useful way of coping with a large difference in levels), the gradient between the pools should be very gentle, with perhaps a waterfall or cascade leading to the next stretch of stream. The width of the stream should also be such that it will appear full when the pump is running.

Overall, the course followed by streams should be simple and uncomplicated. One that is tight and tortuous usually fails to live up to expectations because of difficulties in matching pump outflow with contours.

Other Running Water Features
Waterfalls should be designed as if they arise naturally from the surrounding landscape. They are therefore better suited to water gardens on gently sloping ground. They can be installed in level-ground designs, but this involves additional work to create the high ground from which the water drop will emerge.

As with streams, the width of the lip of the waterfall must be sufficiently wide to take the water flow comfortably; in other words, neither so wide that only a trickle is produced, nor so narrow that this results in a jet of water. Similar considerations apply to cascades, although in their case the water travels from the header pool to the main pond in a series of steps, rather than one single one.

Header Pools
Header pools are small pools constructed at the top of streams, waterfalls and cascades. They are designed to act as a reservoir, from which water will spill over into the desired feature.

Header pools are usually disguised by sinking them into the ground and surrounding them with plants. Filling the pool container with pebbles can help create the illusion that the water flowing into it is actually arising from a natural spring rather than by the artificial means of a water pipe.

Pools that are introduced to break up the flow of a stream are not header pools as such, but still perform a similar function. The front section of these small pools should be made shallower than the back, so that water will remain in them when the pump is switched off – an empty pool is likely to look unattractive, even unnatural.

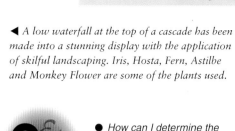

◀ *A low waterfall at the top of a cascade has been made into a stunning display with the application of skilful landscaping. Iris, Hosta, Fern, Astilbe and Monkey Flower are some of the plants used.*

● *How can I determine the correct pump flow rate for streams, waterfalls and cascades?*

To obtain an approximate desired flow rate without having to install the feature first and test it out, allow about 50–60 gallons (c.230–270 litres; c.60–72 US gallons) of water per hour per 1in (2.5cm) of waterfall, cascade or stream lip (the lip is the ledge over which the water flows). A more accurate figure can be obtained using either of the following methods. First, run a garden hose into the header pool and allow it to overflow over the lip. Adjust the flow until a desired overall effect is achieved. Then either: (a) run the water at this rate into a container for 1 minute, measure the volume and multiply it by 60 to give you the hourly flow rate required from the pump; or (b) run the water into a container of known volume and measure the time it takes to produce the required volume. Work out the hourly flow rate from this figure. The figure obtained using either of these methods represents the flow rate required at the head of the stream, waterfall or cascade. This vertical difference between pump and outflow is known as the "head of water" and must be checked when buying the pump.

● *Does the flowing or splashing water produced by running water features affect pond plants in any way?*

While many submerged plants will not be affected by running water, most floating and some surface plants do not like being in the immediate zone of turbulence created by streams, waterfalls and cascades. Floating plants tend to be dislodged by the currents, often resulting in unnatural and occasionally unsightly accumulations, particularly in small ponds. Some surface plants, most notably water lilies, are still-water plants and will therefore be directly and adversely affected by water splashes and currents, and should therefore be located as far away as possible from any source of turbulence.

Siting a Pond

THERE IS NO PRECISE FORMULA TO help you situate your pond in the ideal location. Yet even if there were, finding a real site that met all the requirements would be a difficult, if not impossible, task. As a result, the vast majority of ponds are located in places that represent the best working compromise that prevailing conditions allow.

Siting a pond to its best advantage requires a great deal of thought, both about the positive aspects and about possible pitfalls. Careful forward planning is essential, as mistakes are very difficult and expensive to rectify later. The accompanying checklist will help you to identify potentially suitable sites, and to avoid costly and frustrating mistakes. As well as the key points outlined there, the following are also important factors to bear in mind:

Ease of maintenance will play a major part in your choice of a site. If you are intending to run pumps and filters in the pond, consider where the nearest electrical supply point will be. Remember to allow all-round accessibility both to the pond itself and to the surrounding areas.

Safety at all times must be paramount, as any water feature is a potential hazard, especially to children. Choose a site with all-round visibility, so that you can keep an eye on children while they are playing in the vicinity of the pond.

Sunlight and shade in the correct balance will allow your pond plants and wildlife to thrive. If you live outside northern temperate zones, nearer the tropics, it may be advisable to choose a site that will receive some shade during part of the day. Note that specially constructed trellises or pergolas draped with plants can bring their own problems if the plants are deciduous and will drop leaves on to the pond (see Autumn Maintenance, pages 197–199).

Sloping sites are tricky to plan and construct, but do have aesthetic advantages. Retaining walls or barriers have to be built, either to prevent soil from rolling into the pond, or to hold in the pond water. This extra effort is repaid by the fact that a pond properly terraced into a sloping site looks particularly attractive.

A useful method for assessing the overall suitability of your selected pond site is to mark out the rough shape and size of the pond on the chosen spot with a piece of wide tape or a collapsible garden hose and live with it for a few days. During this time, you will be able to consider all the factors that come into play, such as the amount and duration of sunshine and shade, and accessibility.

▼ *This small garden pond has been thoughtfully sited. The low, open area (centre right) allows clear visibility from across the lawn to the colourful stand of plants around the back of the pond. It also gives easy access for regular maintenance.*

Q & A

● *Why are damp or waterlogged sites unsuitable?*

... While it may sound logical to site a pond in the wettest part of a garden, there are several reasons why such locations are unsuitable. When it rains, the water level in the soil itself (the "water table") rises, generating enormous upward pressure. This can be sufficient to crack concrete, cause large "balloons" inside lined ponds, or even lift prefabricated ponds. In addition, damp sites are particularly difficult to excavate and control.

● *What is the minimum safe distance between a tree and a pond?*

Allow a distance roughly equivalent to the height of the tree, but in the case of some types, e.g. conifers and species such as poplars, which have erect branches that grow close to the trunk, it may be possible to reduce this. A narrow trial excavation should reveal the presence of any main roots that are likely to cause damage. Appropriate root pruning by a specialist may be necessary in cases where no suitably distant pond site is available. Such considerations apply mainly to sunken pond designs. Semi-sunken designs can usually be installed above root level, while fully-raised designs usually only have their foundations below soil level, but well above the roots of surrounding trees.

● *How can I test for unsightly reflections?*

Lay a large mirror on the chosen pond site and view it from all the main vantage points. Placing the mirror at different spots within the pond site will give an accurate picture of how attractive or unsightly most reflections are likely to be.

✔ DO'S AND DON'TS OF SITING A POND

✔ Choose level or only gently sloping ground.

✔ Make sure that your chosen location is low-lying, but above the water table.

✔ Site the pond close to the house and a handy power supply to run the pumps and filters.

✔ Ensure that the pond can be seen from several different vantage points.

✔ In northern, temperate zones, a south-facing location is ideal (a north-facing site will not provide as much light or warmth).

✔ Site the pond well away from trees. Roots can cause damage. Some trees (laburnum, willow) have toxic leaves or fruits, while others (plum, cherry) host water lily aphids in winter.

✗ Avoid wind-prone spots; strong prevailing winds can harm delicate plant shoots. If no other site is available, provide a windbreak.

✗ Don't site the pond in an awkward corner of your garden; allow access all around it.

✗ Sites that lie above electricity cables, or gas pipes should be rejected.

✗ Don't dig your pond in soft or crumbly soil; always select firm ground to stop subsidence.

✗ Avoid any unsightly surroundings that cast ugly reflections on the water.

✗ Don't choose a waterlogged location; this may become flooded during wet spells, and can act as a frost pocket in cold regions.

Pond Design

A GOOD WAY OF GATHERING IDEAS FOR THE design of your pond is to visit aquatic stores or garden centres. Specialist suppliers will have several complete, working set-ups, as well as a range of prefabricated ponds and expert advice at hand. It is also worth seeing public gardens and the schemes of water garden enthusiasts.

Irrespective of whether a pond is intended for a mixed fish collection, Koi, or as a wildlife wetland, they should all have one characteristic in common: a simple, open design. Intricate shapes may look impressive on paper, but do not transfer well into reality. Their narrow inlets and complex bends make them very difficult to manage, as they create still or stagnant backwaters.

Planning a Pond

• Prepare several plans, each incorporating a pond of a different shape and size, located at different points within your general garden scheme. Remember to weigh up carefully the positive and negative factors outlined in Siting a Pond, pages 28–29.

• Give yourself time to play around with your ideas. If you don't want to prepare a new garden plan for each design, simply produce one and then move cut-outs of different pond shapes and sizes within the selected site until you are reasonably happy with a particular arrangement. Alternatively, one of the specialist garden-design programs now available for personal computers will offer you great versatility in drafting a plan.

• Use a tape or collapsible garden hose to mark out on the ground both the shape and size of the pond design you have opted for on paper. (You can use small stakes or chalk, though this will not allow you to envisage the final effect as clearly.) Live with this design for at least several days, modifying it as new ideas emerge, until you are satisfied that you have something that looks attractive, fits well into your garden or patio setting, and will meet all the basic requirements of the fish and plant collections you have in mind.

Open designs need not be restricted to squares, circles or kidney shapes, attractive though these are. There is a huge variety of designs that still adhere to the principle of being open and simple.

Formal and Informal Designs

Pond designs are usually categorized as either formal or informal. Formal designs are hallmarked by symmetry, or "controlled" asymmetry, in their overall shape. Typical symmetrical examples are square, circular, matching straight or curved-edge designs. If ponds are not entirely symmetrical, but nevertheless consist of sections that are straight-sided or smoothly curved, their controlled asymmetry still places them in the formal category.

Informal layouts all depart to some extent from such formality. The best informal designs create the illusion that they have been formed by natural forces. The truth is usually quite different, in as much as the most attractive informal ponds are generally those that have been most meticulously planned in advance. In practice, most designs fall somewhere between the two extremes of a square, rectangular, circular or ovoid formal shape and the more or less random informality of a genuine wildlife pond.

It is important not to consider your chosen pond design in isolation, but as an integral part of the overall garden or patio scheme. If the garden is wider than it is deep, installing a pond that is deeper than wide can make the overall arrangement look awkward. The same is true of wide ponds in layouts that extend in a long, narrow strip away from the house. Square sites can take a broader range of shapes, and siting the pond near to one of the far corners rather than the centre of the garden will enhance layouts.

In deciding whether to have a formal or informal design, consider its compatibility with your garden style. Above all, avoid siting a formal pond in a wildlife garden, or vice versa, since they will detract from one another.

● *What are the relative merits of raised and sunken pond designs?*

... Raised designs are unsuitable for wildlife schemes, as they prevent easy access to and exit from the water. They also look too unnatural for this purpose. On the other hand, they are ideal for deep-water designs, such as Koi pools (see pages 12–15), giving you adequate water depth while sparing you a lot of digging. Raised designs are safer for children – although children should *never* be left unsupervised, whatever the pond type. Raised designs are also better for the elderly and physically impaired, especially if spaces are left around the edge for wheelchair access. Sunken ponds look particularly comfortable within informal designs, but are more prone to wind-blown debris than raised ponds. Conversely, they are less susceptible to environmental fluctuations than raised designs, although if a raised pond is large enough, these fluctuations are not likely to be excessive, except under severe weather conditions.

▲ *A formal pond on a patio. This scheme offsets the straight-sided symmetry of the pond with the topiary conifers in the foreground and the ivy spiralling around the statue in the background.*

▶ OVERLEAF *An informal water garden design, incorporating a shallow pebble beach to allow amphibians easy access to the pond.*

● *Can a pond be designed with vertical sides?*

Ponds should have sloping sides (about 70°) in areas where the water surface freezes in winter. Theoretically, the slope will cause freezing water, which expands and exerts considerable force, to slide upwards, rather than press against the walls. In practice, though, whatever the gradient of the walls, as long as they are sturdy enough, freezing water always tends to rise rather than crack the pond. However, to be absolutely sure that a pond will withstand even the lowest temperatures, it is wise to incorporate sloping sides in the design.

Prefabricated Ponds

PREFABRICATED PONDS, ALSO REFERRED TO AS rigid or moulded units, have the advantage of being generally easy to install, and many people like them because the shape is predetermined. They usually come with built-in shelves.

When they first appeared on the market, most prefabricated ponds were small and shallow. While these are still available – and are adequate for many garden and patio schemes – deeper and larger units have become ever more popular. It is now possible to buy a prefabricated pond "off the shelf" that can accommodate anything from a shoal of minnows to a complete Koi collection.

Provided they are sufficiently deep, prefabricated ponds are especially suited to formal layouts, where accurate, matching pond dimensions are required and where the time or expertise necessary to construct such a pond is not available. Moreover, prefabricated ponds are essential if the design needs to be crease-free, since this is virtually impossible to achieve using a liner.

As well as coming in a variety of shapes and sizes, prefabricated ponds can also be bought in a range of materials, from vacuum-moulded plastic designs to more durable fibreglass, PVC and rubberized compounds. Prices tend to reflect durability, a vitally important factor when permanent installations are being considered.

One other factor that you should keep in mind when choosing a pond of this type is its colour. As a general rule, darker colours tend to blend in more naturally with most patio and garden settings, but lighter-coloured ponds can sometimes look very impressive. Bright colours – which are still available on some of the smaller prefabricated units – are usually best avoided.

Pond Edging

Before making a final decision on what type of prefabricated unit to buy, you should consider the nature and design of the pond surround.

Most prefabricated ponds have a lip all round the edge, which can measure 4in (10cm) wide or more. Such lips are difficult to hide using vegetation or other informal means, but can be easily and instantly covered up with attractive paving slabs. If you prefer to plant around the edges, it is worth searching for a narrow-lipped pond design, but this is likely to restrict your choice to the more expensive models.

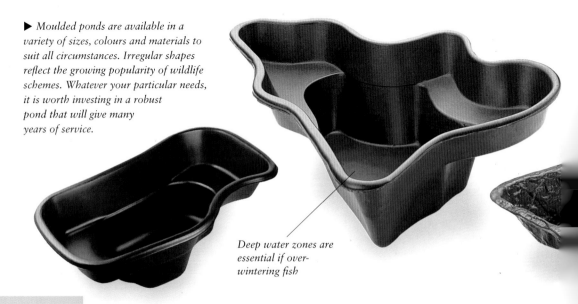

▶ Moulded ponds are available in a variety of sizes, colours and materials to suit all circumstances. Irregular shapes reflect the growing popularity of wildlife schemes. Whatever your particular needs, it is worth investing in a robust pond that will give many years of service.

Deep water zones are essential if over-wintering fish

▲ *A prefabricated pond provides a convenient way of achieving a formal look. This classic circular design is complemented by paving and a neat lawn. Units are available in bright shades, but these can look garish.*

Marginal shelves
for plants

Rough "rockwork"
effect

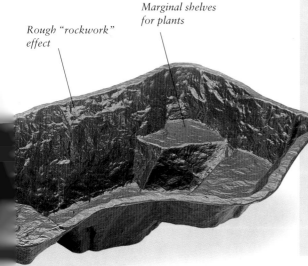

Choosing a Prefabricated Pond

• Visit an aquatic centre with an extensive selection, including working displays.

• If this is not possible, check the supplier's catalogue thoroughly and obtain as much information as possible.

• Write down the dimensions and mark out the pond outline on your site with a garden hose (seen in shops or catalogues, ponds appear larger than they actually are).

• Go for the largest affordable model that the site can comfortably accommodate.

• Check that the shelves will be able to accommodate all the marginal plants you have in mind.

• Consider the more durable materials if you are planning a permanent installation.

• If you are intending to stock the pond with fish, check that the chosen design will provide adequate water depth (a minimum of around 18in/45cm for areas that experience either very cold winters or summer heatwaves).

DIGGING OUT

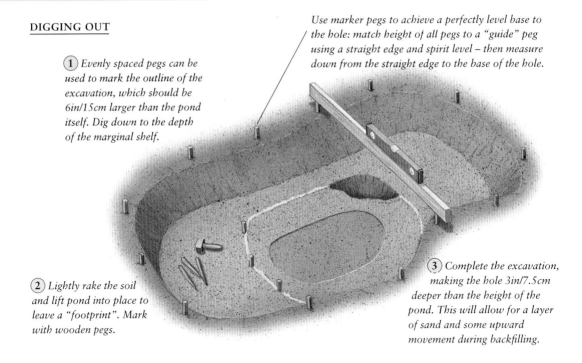

Use marker pegs to achieve a perfectly level base to the hole: match height of all pegs to a "guide" peg using a straight edge and spirit level – then measure down from the straight edge to the base of the hole.

(1) *Evenly spaced pegs can be used to mark the outline of the excavation, which should be 6in/15cm larger than the pond itself. Dig down to the depth of the marginal shelf.*

(2) *Lightly rake the soil and lift pond into place to leave a "footprint". Mark with wooden pegs.*

(3) *Complete the excavation, making the hole 3in/7.5cm deeper than the height of the pond. This will allow for a layer of sand and some upward movement during backfilling.*

Installing a Sunken Prefabricated Pond

To mark out a symmetrical pond, turn it upside down and draw the outline accurately on the ground. (Irregular or informal designs cannot be marked out like this; the resulting outline will be reversed.) If the pond is irregularly shaped, place it upright and level it out with supports so that a series of stakes, straight sticks, canes, or rods can be driven vertically into the ground all the way around the pond outline. This gives an approximate shape that can be further refined by looping a tape or garden hose at ground level around these guides.

For an easier outline, measure the maximum length of the pond, and the width at several points along the sides. Mark out an approximate rectangular shape on the ground slightly larger than these dimensions. This method does mean having to dig more earth, though.

Excavating the Hole Remove the pond from the marked outline, measure the depth and size of the shelf or shelves and mark their overall size and position within the pond outline. Dig the whole pond to just below this level, making sure that the eventual excavation is some 4–6in (10–15cm) wider and longer than the pond itself. Rake over the base of the hole, stand the pond in it and press down to produce an outline of this deeper section on the excavated surface.

Remove the pond and excavate the newly marked section down to about 3in (7.5cm) below the bottom of the pond. Keep the horizontal bottom surface level by measuring the depth vertically from a level pole or strip of wood laid across marker pegs at the top of the excavation.

Check the excavation thoroughly, remove all sharp stones, compact the soil and spread a 2in (5cm) layer of builder's sand on the base. Lower the pond gently into position and adjust the excavation if necessary. If you have measured

You will need...

- Builder's sand or sifted stone-free soil
- Garden canes and wooden stakes (12in/30cm)
- Tape measure and string
- Spade, rake and other gardening tools including hose pipe, hammer
- Spirit level and wooden straight edge
- Bricks and blocks to support pool during marking out
- Tamping tool, such as a pole or a piece of wood, with a smooth, flat end.

INSTALLING

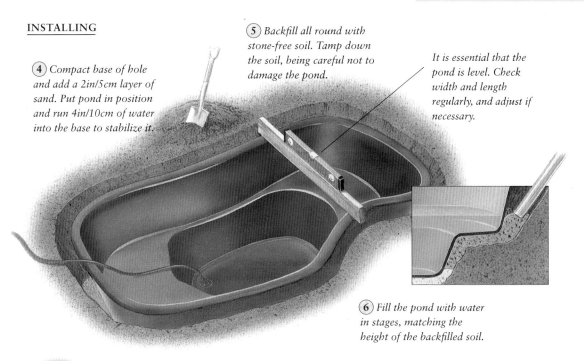

④ *Compact base of hole and add a 2in/5cm layer of sand. Put pond in position and run 4in/10cm of water into the base to stabilize it.*

⑤ *Backfill all round with stone-free soil. Tamp down the soil, being careful not to damage the pond.*

It is essential that the pond is level. Check width and length regularly, and adjust if necessary.

⑥ *Fill the pond with water in stages, matching the height of the backfilled soil.*

Q&A

● *Is it possible to install raised prefabricated ponds?*

... Yes. Pond moulds are usually partially or semi-raised anyway, with the pond being sunk in the ground to the level of the first shelf. This sets the pond on a firm footing with soil support all along the bottom of both the pond itself and the shelves. The above-ground part of the pond also requires support. This can be provided in several ways, for example by building a surrounding wall (or using a pond surround such as a log roll) from ground level up to the underside of the pond lip. Alternatively, construct the wall (or equivalent) slightly farther away and backfill the resulting gap with firmly compacted sand or soil. During this installation, use a spirit level laid on a straight-edged wooden strip extended across the pond, and check it frequently to ensure that the pond is perfectly aligned with the horizontal.

● *Do all prefabricated ponds have shelves?*

Most do, and some have more than one. Designs without shelves are less common, but they do exist.

● *What is a sectional pond?*

Sectional ponds come in a series of modules which can be fitted together to create a customized design out of prefabricated units. Full instructions on how to assemble modules, along with the fittings and sealant required, are provided by the manufacturers.

accurately and worked carefully, the lip of the pond should end up lying just below ground level and need little adjustment.

Backfilling and Installing When the pond is firmly seated and level, run water into the base to a depth of 4–6in (10–15cm). This gives stability by bedding the mould into the excavation. In addition (particularly with less robust models), the outward pressure exerted by the water counteracts the inward pressure produced by compacting the backfill. The risk of the walls buckling is minimized by gradually increasing the water level as backfilling proceeds.

Now backfill around the pond with sand or stone-free fine soil up to the water level and tamp it down. Repeat, using a spirit level to check, until the backfill is several inches below lip level. Plants can be placed in position now or after the pond has been decorated inside. If you are paving the pond edge, create a trench at least as wide as the slabs and measuring about 4–6in (10–15cm); dig it all round the pond and fill in with either compacted hardcore or a cement/sand/gravel mix. Fill the underside of the lip with well-compacted sand. Once the mix has set, lay the slabs with a standard cement/sand mix, overlapping the pond edge to hide the mould from view.

Lined Ponds

WHEREAS PREFABRICATED PONDS HAVE THEIR shape predetermined, lined ponds require more planning. Every detail, from the size of the excavation to the slope of the sides, the pond outline, and the number, distribution, shape and size of shelves, must be fixed before digging begins. Each lined pond is therefore "customized" to your personal preference, needs and circumstances.

Lined ponds are ideal for informal schemes, since the materials from which they are made are pliable, and so will fit most shapes and contours, albeit with varying amounts of creasing. Liners can be used just as effectively in formal designs, although the wide range of formal prefabricated models available makes building this type of lined pond a less attractive option.

Liners are usually associated with sunken designs, but they need not be restricted to this type of pond. In fact, many raised ponds which at first sight appear to be constructed entirely of bricks, cement, blocks or even logs, are actually lined inside. Such arrangements take advantage of one of the many positive aspects of liners: their instant, effective and flexible waterproofing.

Types of Liners

Liners come in a variety of materials and prices: **Polythene** is the most common material for cheaper, less permanent liners. It lacks pliability and becomes brittle after being exposed to the sun's ultra-violet radiation. Unless you are planning a temporary set-up, polythene is best avoided.

Butyl rubber, PVC, and LDPE (low-density polyethylene) are used for many of the more expensive types of liner. Butyl and PVC liners can be obtained in various "grades" (thicknesses); some of the latter incorporate a nylon/terylene web for added strength. Among the rubberized liners, some types have a layer of pebbles glued to the top surface, which lends the exposed strip around the edge of the pond the appearance of a natural "shore". PVC and butyl liners have varying degrees of elasticity and resistance to sunlight, and are guaranteed for between 10 and 50 years.

Clay-impregnated textiles (or "geotextiles") are now available as liners for wildlife pond enthusiasts who would like a clay-lined pond without the hard work involved in puddling (see Wildlife Ponds, page 16). Impregnated with sodium bentonite (clay), they are "self-healing"; in other words, if they sustain a minor puncture, the bentonite will plug the hole.

Underlays

For a lined pond, it is also wise to invest in an underlay, which acts as a cushioning material between the liner and the surface of the excavation. A range of manufactured underlays is now widely available; these are without doubt the most convenient way of creating a protective lining. However, if underlays cannot be obtained, or if you want to use an alternative, then any non-toxic material – for example sand, old carpets, fibreglass insulating felt, or polythene bags – can be used. Even thick layers of newspaper will do, but these will obviously not last as long.

Choosing a Pond Liner

- Buy the best liner you can. Good-quality standard-gauge liners will do for most ponds, but check your needs at an aquatic store.
- Don't consider a liner in isolation; give serious thought to the underlay as well and buy at least as much underlay as liner.
- Don't buy either liner or underlay until the size of the excavation is accurately known.
- Consider a bentonite-impregnated liner for a wildlife pond; they are the most natural-looking manufactured liner.
- Don't skimp on size; always allow a generous surplus. It is far better to have too much liner than too little.

● *How do I work out how much liner I am going to need?*

... This calculation is much easier than it might appear. Follow these simple rules: (a) Measure the maximum length, width and depth of the pond. (b) Add the maximum length and twice the maximum depth (the liner has to extend down both ends of the pond to the bottom). The resulting figure gives you the length of liner required; (c) Do the same calculation for the width (i.e. width plus twice the depth). This will result in a liner that fits the pond, but with little safety margin. Adding some 18–24in (45–60cm) to both the length and width calculations should also provide a reasonable amount of overlap around the edges.

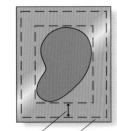

Pond depth   "Extra" for edging

◀ *A liner forms the basis of this informal pond scheme; it has been concealed using a variety of plants and edging rockwork.*

▼ *Different liners and underlays. Liners: 1 blue woven LDPE; 2 black PVC; 3 green reinforced PVC; 4 butyl rubber. Underlays: 5 old carpet; 6 underlay felt.*

1   2   3   4   5   6

Installing a Lined Pond

First, go round the outside of the outline of the pond, marking it with a spade and removing a thin layer of turf, at least as wide as the spade itself, but preferably wider. This strip will eventually form the edge of the pond, onto which the liner will overlap and over which the slabs or other type of pond edging will be installed.

Excavate the hole, sloping the sides at about 70° (except in wildlife schemes, where the slope should be very gentle). Even in other types of design, it is a good idea to create at least one gently sloping area to afford easy access and exit for wildlife. Shape the shelves as you dig down, allowing enough width for planting baskets (12in/30cm should be regarded as the bare minimum). Shelves can be incorporated at different depths to cater for a variety of plants, except in saucer-shaped wildlife designs.

If the soil is very crumbly, it may be necessary to install a supporting back wall to the top shelf to prevent damage to the edge once the pond is in use. This can be done by lining the back of the shelf with building blocks or bricks extending from the horizontal surface of the shelf to what will be the eventual water level. The retaining wall should be cemented in place using a standard 3:1 cement mix.

You will need...
- Builder's sand or underlay material
- Small wooden stakes (12in/30cm)
- Tape measure and string
- Spade, rake and other gardening tools including hosepipe and mallet
- Spirit level and wooden straight edge
- Bricks/blocks to weigh down the edges
- Edging material of your choice

As excavation proceeds, keep checking with a spirit level to ensure that all the shelves are horizontal; it is not so vital for the bottom of the pond to be absolutely level. Dig a few centimetres below the required water depth to allow for the thickness of the underlay if this is going to consist of a thick layer of sand; this additional depth will not be so necessary if a commercial underlay or carpet is being used.

After removing any sharp stones, lay the underlay, pressing it down firmly and ensuring that it overlaps the top margin, especially if this consists of sharp-edged bricks or blocks. If you are using sand, dampen it slightly for easier handling. With the underlay in place, there are several ways of installing the liner. For example:

DIGGING AND LINING THE HOLE

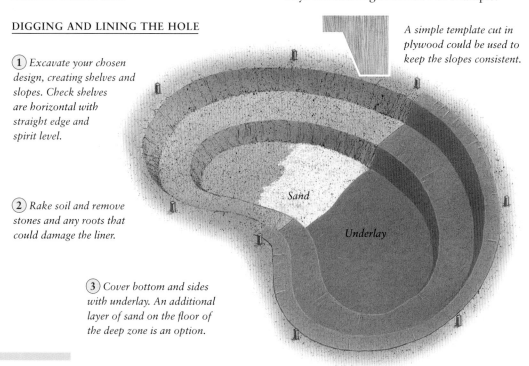

1 *Excavate your chosen design, creating shelves and slopes. Check shelves are horizontal with straight edge and spirit level.*

2 *Rake soil and remove stones and any roots that could damage the liner.*

3 *Cover bottom and sides with underlay. An additional layer of sand on the floor of the deep zone is an option.*

A simple template cut in plywood could be used to keep the slopes consistent.

Sand

Underlay

(i) Mould the liner into position in the pond, smoothing out as many creases as possible. Run a hose into the deepest part of the pond and adjust any folds that occur as the pond fills up.

(ii) Alternatively, stretch the liner over the excavation, allowing it to sag slightly and weighing down the edges with bricks, blocks or smooth-edged rocks. Fill slowly, and adjust creases as they arise; the weight of water will gradually mould the liner into position; to minimize wrinkling, maintain tension by moving the bricks.

Once the pond is full, remove the weights and leave the liner to settle and mould itself into its final position. Then trim the edges of the liner, leaving sufficient width all round to form a good overlap with the paving or other pond surround.

If the surrounding terrain is firm and paving slabs are being used, these can be cemented directly on top of the liner. Place them to overlap the edge of the pond and hide the liner. If the terrain is crumbly, dig a trench about 8–12in (20–30cm) wide and 4in (10cm) deep all round the pond and fill it with concrete to the same level as the top edge of the shelf-supporting wall. Once this has set, it can act as the base for cementing the paving slabs.

Q & A...

● *Should liners be treated or handled in any special way to make them easier to install?*

Warm temperatures will make liners more flexible and easier to handle. So, if possible, unroll the liner close to the pond and leave it in the sun for about 30 minutes. If there is no sun, or if the ambient temperature is low, place the liner in a warm room for several hours (but not in direct contact with a heat source). Just prior to installation, roll the liner up again and lay it at right angles to the long axis of the pond. Place some weights on the free end and slowly unroll the liner as you move down the side of the pond (this is a two-person job). Try to keep in line with the person on the opposite side of the pond, holding your body as upright as possible and the liner roll at waist height. Alternatively, to make sure of centring the liner over the pond, roll the liner up from each end to make two rolls meeting in the middle. Lay this double roll across the middle of the hole, and then unroll one side at a time.

◀ ▼ *Once the pond liner has been installed, there are a number of different strategies that you can use to conceal its edge. These include (left) cementing paving slabs, or (below) laying turf or edging stones, or creating a marsh area.*

INSTALLING THE LINER

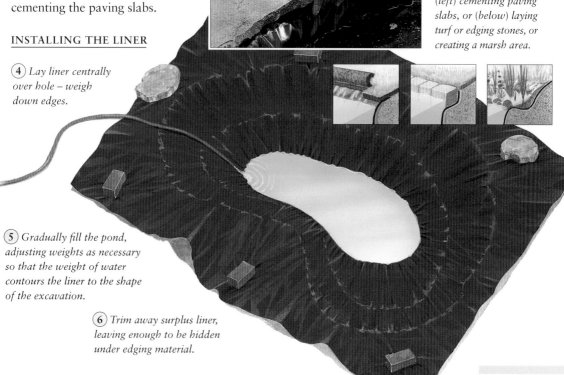

④ *Lay liner centrally over hole – weigh down edges.*

⑤ *Gradually fill the pond, adjusting weights as necessary so that the weight of water contours the liner to the shape of the excavation.*

⑥ *Trim away surplus liner, leaving enough to be hidden under edging material.*

Concrete Ponds

IF YOU HAVE NEVER WORKED WITH CEMENT before, the idea of building a concrete pond may appear daunting, and with good reason! Getting the right mix, applying it correctly, and keeping it workable – all this may well deter a beginner. Yet properly designed and built concrete ponds have an air of permanence and elegance that most other construction materials cannot match.

Materials

Concrete can be bought either ready-mixed, or mixed at home from the basic constituents of cement, sand (or gravel) and water. Especially if you are inexperienced, consider carefully in advance which sort you should use.

It may sound an attractive option to order ready-mixed concrete, but unless you are able to work with it immediately, it may begin to harden before you have completed the job. Mixing the concrete or mortar yourself may be a better alternative, as you will be able to work in manageable batches. Make sure that each batch of mix is identical to the last one, so as to maintain uniform properties throughout the pond.

Waterproofing compounds can be used to enhance the water-retaining properties of the main concrete mix. Alternatively, you can lay an impervious sheet (e.g. heavy-gauge builder's polythene) as an excavation lining, onto which you then pour the concrete. To help prevent cracking in the final layer, reinforcing fibres can be added.

Constructing a Concrete Pond

The following is a basic guide to building a pond out of poured concrete.

First, decide on the pond outline, using a tape or hose (see Pond Design, pages 30–31). When marking the size of the final excavation, remember to factor in the width of the walls (for concrete, usually about 8in/20cm), and also to make allowances for any shelves.

After digging the hole, install a hardcore layer measuring 4–6in (10–15cm) along the base and cover it with a concrete layer about 4in (10cm) thick, which you also spread up the sloping sides and on the shelves. While this is still wet, press a reinforcing wire mesh or web (ideally covering the whole pond) into the surface of the concrete and top it with another 2in (5cm) concrete layer. Once the concrete has set overnight, render the surface with a thin layer of mortar (sand/cement mix), with or without reinforcing fibres. Make sure that you end up with a smooth, even finish.

After two days, it will be safe to step into the pond. The fine dust from the surface of the rendering should now be brushed off and the pond walls and base painted with one or two coats of a proprietary sealant. This not only improves waterproofing but – most importantly – prevents toxic cement lime from leaking into the water.

You will need...

Materials
- Concrete 1 part cement:2 parts sand:
  4 parts coarse gravel
  (+ Waterproofing compound)
- Mortar rendering
  1 part cement:2 parts sand:
  Reinforcing fibres at the rate of 25–30g/1kg of dry sand/cement mix
- Wire mesh (2in/5cm)
- Bricklayer's sand and/or hardcore if the subsoil has a high clay content
- Sealant or coloured pond paint

Tools and Equipment
- Waterproof gloves
- Watering can with fine rose
- Plasterer's "float" (rectangular trowel) to smooth mortar
- Spirit level
- Sacking, to keep concrete damp
- Boards to work from
- Shovel and gardening tools
- Cement mixer, if the pond is large.

CONSTRUCTING A CONCRETE POND

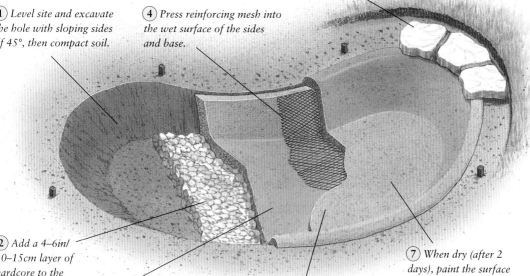

6 *Edge pond in an eye-catching way. Overhanging edging stones are shown here.*

1 *Level site and excavate the hole with sloping sides of 45°, then compact soil.*

4 *Press reinforcing mesh into the wet surface of the sides and base.*

2 *Add a 4–6in/ 10–15cm layer of hardcore to the pond base.*

3 *Cover base and sides with a 4in/10cm layer of concrete. Keep moist with damp sacking.*

5 *Lay additional 2in/5cm layer of concrete over the top of the mesh.*

7 *When dry (after 2 days), paint the surface with a sealant or coloured pond paint.*

Ponds with vertical, rather than sloping, walls are far more difficult to install, as wooden shuttering must be erected to keep the poured concrete in place until it sets. Both shuttering and the insertion of the reinforcing wire mesh into the wet concrete walls demand considerable skill. Therefore, unless you are experienced, it is best to seek help from a professional builder. Ponds with vertical walls can also be built from bricks or cinder blocks, cemented together and finished with a layer of rendering that incorporates fibre reinforcement.

Concrete pools for Koi are even more complex. Deeper than conventional designs, they may have reinforcing rods in the base and walls, and usually include one or more drains and their plumbing. They may also be lined with fibreglass. Get advice from Koi keepers, or from specialist literature (see Further Reading, page 202).

The pond should be finished off with your chosen edging material to conceal the concrete rim. Remove any fragments of cement mix or other bedding material that fall into the pond before they have a chance to set.

● *Should concrete be protected during the drying/curing stages?*

... Yes, in freezing, very wet, or hot and dry weather. Extreme cold should be avoided, since it makes the materials difficult to handle and affects the drying process. Although the mix can be made more pliable with a small amount of liquid detergent or antifreeze, these should not be used in any structures, such as pond walls or bases, that will eventually lie underwater. In wet conditions, cover the concrete with plastic sheets to prevent it from being washed away. In dry heat, cover it with damp sacking to stop it from drying too quickly and cracking.

● *How can the toxic effects of lime in the concrete be neutralized before the pond is stocked?*

If the pond has been properly sealed, no toxins should leach out. To be sure, use a pH test kit after filling to ascertain the water chemistry – over pH8.5 indicates a high lime content. To neutralize it, add enough potassium permanganate to the pond to create a light pink/purplish colour. Alternatively, use hydrochloric or acetic acid at a rate of 1ml per 100 litres (22 gallons/26 US gallons) of pond volume. Rinse thoroughly, then refill and re-test the pond before stocking it with fish.

Running a Pond

A POND IS ESSENTIALLY A SMALL, ENCLOSED body of water that is, in many respects, cut off from the outside world. Although it is affected by climate, falling leaves, dust, insects and many other factors, it lacks a number of important ingredients that most healthy, natural bodies of water possess. These include a constant supply of fresh water, continual removal of "old" water, a large bacterial population to process wastes, and a food supply that is in balance with the pond inhabitants.

If an artificial pond can successfully reproduce its natural counterparts by being large enough and sparsely stocked with fish, and by containing abundant plant growth, then it will become almost self-sustaining. Yet this is the exception rather than the rule; the majority of ponds constructed in gardens or on patios tend to contain a somewhat higher ratio of fish to plants than an equivalent natural body of water. In such circumstances, conditions in a pond will quickly deteriorate to an intolerable level, unless artificial means are employed to prevent this from happening.

▲ A submersible pump can be used to power supplementary pond water features, such as this impressive fountain.

SUBMERSIBLE PUMPS

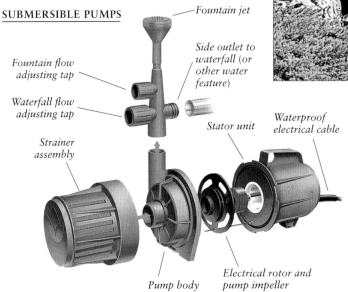

Fountain jet

Fountain flow adjusting tap

Side outlet to waterfall (or other water feature)

Waterfall flow adjusting tap

Strainer assembly

Stator unit

Waterproof electrical cable

Pump body

Electrical rotor and pump impeller

◀ Diagram showing the constituent parts of a submersible pond pump. Convenient units such as this have the advantage over surface pumps that they run silently, without emitting a constant hum.

The two most important pieces of equipment in helping an artificial pond to sustain life are pumps and filters. In recent years, ultra-violet (UV) sterilizers or clarifiers have also become widespread; these are certainly valuable aids to water-quality control, but are not absolutely essential to running a pond.

Pond Pumps

Despite the apparently bewildering range of pumps on the market, they are all basically of two main types: submersible and surface.

Submersible pumps, as their name indicates, are designed to operate while fully submerged, and will burn out if run above water. Submerged pumps are generally more suitable for small and medium-sized ponds and (depending on their output) can be used to run fountains, waterfalls, filters, cascades and streams in such systems.

Surface pumps are designed strictly for "out-of-pond" use and will suffer irreparable damage if any attempt is made to operate them underwater. They are usually more powerful than submersible types and are therefore suitable for larger displays, or where the motor is required to move considerable volumes of water or push it over long distances.

The following safety precautions should always be observed when installing a pond pump. First, make sure that surface pumps will not be

● *What advantages are there to using a low-voltage pump?*

... Low voltage does not mean low performance, at least not with small or medium-sized pump models. Nor does the fact that they are cheaper than their high-voltage counterparts signify inferior quality. Thus, on both these counts, there is little to choose between the two types. Where major systems are concerned, however, i.e. those where large volumes of water need to be moved, low-voltage pumps are generally not available. The most significant advantage that low-voltage models have over mains-operated pumps is their safety. The voltage they operate on (usually 24V) will not produce an electrical shock if the cable is cut, whereas mains electricity, whatever the voltage, most certainly will. Although low-voltage pumps require a transformer, they can nevertheless still work out cheaper to install than conventional models which – for safety reasons – should be plugged into a Residual Current Device (RCD) or circuit-breaker.

Choosing a Pump

• Visit a specialist centre and seek advice.

• Most aquatic outlets have a working selection of pumps on display. Assess the relative merits of several models "in action".

• Check the manufacturers' specifications of as many pumps as possible.

• Choose a pump that will be able to move the entire volume of your pond about once every 2–3 hours.

• Check that your chosen pump can raise water in sufficient quantities to the height needed to run special features (see Streams, Waterfalls and Cascades pages 26–27).

• Decide what type of fountain spray (if any) you want, and to what height you would like the water lifted, and check that the pump is capable of achieving this.

▼ *Submersible pumps and (far right) a surface pump. High-output surface pumps are favoured by pond keepers with large ponds or several water features to run.*

immersed and that submersible models will not be exposed to the air once they are in operation. For wiring the pump, all cables, plugs and switches must be suitable for outdoor use, and all connections waterproof. If possible, keep connections above ground level and, for preference, enclosed within appropriate trunking or conduits. Provide each piece of electrical equipment with a separate Residual Current Device (RCD); these act as circuit-breakers in the event of a fault or physical damage to a cable. Finally, always switch off the electricity supply before handling any equipment and, if in doubt, call a qualified electrician.

The best position for a pump largely depends upon the type of pond filter you are using (see below). If the pump is being used in conjunction with an internal filter, then the nearer it is to the filter the better the flow of water through the filter. In the case of an external filter, pumps should be located as far from the outflow of the filter as possible to ensure that all the water in the pond is pumped to the filter on a regular basis. In summer, it is a good idea to raise the pump slightly from the pond bottom to minimize the amount of debris that will be drawn in.

Pond Filters

The aim of fitting a pond filter is to establish and maintain adequately healthy water conditions for the long-term well-being of the pond inhabitants. Like pumps, filters can be installed either

◀ A pump "pre-filter" is a useful piece of equipment for small ponds. Comprised of a series of foam sections, this unit fits onto the inlet pipe of a submersible pump, and restricts the uptake of solid waste that would block the pump.

▲ An external tank filter. The brushes and sponges in such units are easily removable for cleaning. Several tank filters can be linked in series to provide efficient mechanical and biological filtration for larger ponds.

Aids to Filtration

The following are now widely available as useful aids to filtration:

Ultra-violet sterilizers are also (correctly) referred to as clarifiers and (incorrectly) as ultra-violet filters. A filter must incorporate a type of sieve that can physically remove particles from suspension. UV sterilizers do not do this. Instead, they destroy the nucleic material of some free-floating pathogenic (disease-causing) organisms and cause cells of free-floating algae to aggregate together (flocculate) into small clumps, which can then be trapped by a mechanical filter medium.

UV sterilizers/clarifiers thus perform two useful purposes: they help control pathogenic

micro-organisms and also help to eliminate green water. Many UV units are now integral components of filters and have come to be regarded by many as essential pieces of equipment (although they are not).

Magnetic water treatment is a recent innovation to help reduce the limescale deposits that impair the efficiency of filtration units and UV clarifiers. Magnets are fitted between the pump and the filter, and work by altering the structure of carbonates and bicarbonates in the water. One side-effect of magnetic descaling is that it retards plant growth (including nuisance blanket weed). However, such units may also cause fluctuations in the pH of the pond.

FILTRATION OF POND WATER

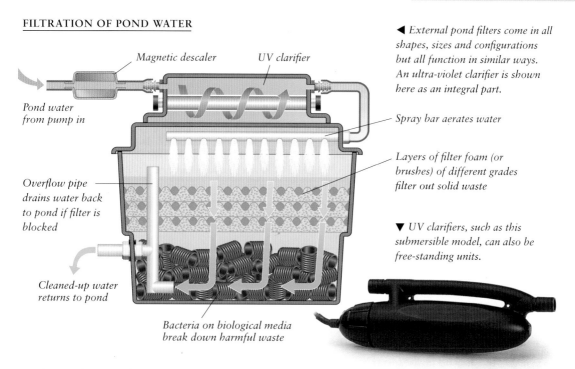

Magnetic descaler

UV clarifier

Pond water
from pump in

Overflow pipe
drains water back
to pond if filter is
blocked

Cleaned-up water
returns to pond

Bacteria on biological media
break down harmful waste

◀ *External pond filters come in all
shapes, sizes and configurations
but all function in similar ways.
An ultra-violet clarifier is shown
here as an integral part.*

Spray bar aerates water

*Layers of filter foam (or
brushes) of different grades
filter out solid waste*

▼ *UV clarifiers, such as this
submersible model, can also be
free-standing units.*

in the pond or outside. Filters also employ a variety of different filter media (see pages 48–49). Before deciding, look through catalogues, and discuss your needs at an aquatic centre or with experienced pond keepers. One key point to bear in mind is that, for a filter to work efficiently, it must be teamed with a pump that supplies it with a sufficiently strong flow of water. Manufacturers recommend that the entire volume of the pond should be passed through the filter approximately every two or three hours.

Internal filters, which are fitted underwater, are generally quite straightforward to install and dismantle. Most models come as self-contained units, complete with a single filter medium or a selection of media and inbuilt inlet and outlet ducts. Some also have integrated pond pumps, with or without fountain attachments.

However, there is another type of internal filter that is designed to be incorporated into the fabric of concrete or block ponds. This must be planned at the same time as the pond itself, since it may require a retaining wall and pipework, and can occupy a large area of the pond bottom (up to a third). Such filters usually consist of a thick layer of gravel through which the dirty pond water is pumped. Debris is removed by the gravel grains and dissolved toxic wastes are neutralized by bacterial action. Incorporating this type of internal filter (more common in Koi pools than in general ponds) into the overall design requires great skill, and should only be undertaken by an experienced person. Fortunately, a simpler version of the inbuilt internal filter is now available, which only requires that the filter medium be contained within a retaining barrier or wall, or, alternatively, spread as a layer over the whole base of the pond. This filter usually takes the form of a set of perforated pipes radiating from a central column, which is buried under the filter medium and connected to a pump.

External filters can either be built into the pond design or bought as complete units. Off-the-shelf models usually incorporate all the necessary pipework and filter media – and often a UV clarifier as well – within a single cistern-type container. The fittings on such units are easy to connect to the servicing pond pump and the pond itself. Building a sophisticated filter into the fabric of the pond is far more complex, and is usually reserved for specialized Koi pools.

Information on other filtration aids, such as vortex chambers, cartridge filters and sand filters, can be found in Koi Pools, pages 14–15.

Filtration and Filter Media

In order to be really effective, a filter system must be able to function in three different ways – mechanically, chemically and biologically. This is achieved by means of filter media, some of which are better suited to performing one type of water treatment than another. In practice, it is not just the medium used, but the way in which it is used, that determines its effectiveness. Of the three types, only mechanical filtration can be regarded as true filtration, in as much as it removes solid particles from the water. The other two processes involve the removal or transformation of dissolved compounds, and should therefore properly be called "purification" or "detoxification" processes.

For maximum effectiveness, media should be kept separate from each other, with mechanical media forming the first layers through which solid-laden pond water is pumped. This can be followed by a chemical medium, if necessary, and then by one or more layers of biological media. Each type of medium can be housed in a separate filter chamber or, as in most filter types, as separate layers within a single chamber.

The following is a selection of media that are most commonly available in aquatic stores and garden centres, although there are others:

Mechanical filtration physically removes solid particles from the water and is a vital component of every filtration system. Mechanical media for filters include brushes, floss, nylon scouring pads, filter matting and foam and Canterbury spar. In most cases, these mechanical filter media will also function biologically, if they are allowed to operate without thorough cleaning for a long time, as their surfaces become colonized by beneficial bacteria. To ensure optimum operation, therefore, mechanical media should be cleaned on a regular basis, preferably by being rinsed in pond water. Failing this, you can use tap water or rainwater. On no account should you *ever* use detergents, however, because even if mechanical media are thoroughly rinsed following detergent treatment, minute residues can remain that are particularly toxic to bacteria.

Chemical filtration involves the removal of certain dissolved toxic wastes such as ammonia and some organic compounds through a process of adsorption, which effectively "sticks" molecules onto the surface of an appropriate filter medium. Chemical media include zeolite chips or granules and activated carbon. Filter media such as these are particularly useful for adsorbing ammonia in new ponds, before the ammonia-processing bacterial population has had a chance to become established.

Exhausted zeolite can be recharged by soaking it in brine. Activated carbon, on the other hand, is best discarded and replaced with an entirely fresh batch.

▼ *Filter media come in all shapes and sizes, a selection of which is shown here. Most of these can function in more than one way, e.g. mechanically and biologically, depending on how they are used.*

MECHANICAL FILTERING MEDIA

CHEMICAL FILTERING MEDIA

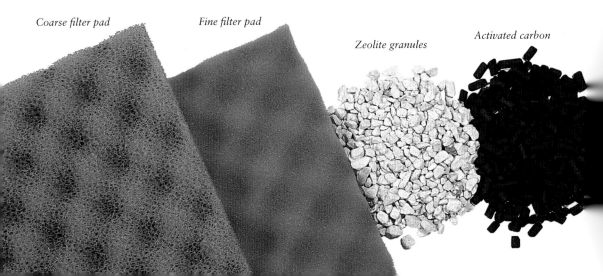

Coarse filter pad

Fine filter pad

Zeolite granules

Activated carbon

Q&A...

● *What is the difference between aerobic and anaerobic filter bacteria?*

Aerobic filter bacteria require dissolved oxygen to perform their water-purifying metabolic processes. Anaerobic bacteria do not.

● *How are nitrification and denitrification different?*

Nitrification is the process by which aerobic bacteria convert toxic ammonia into less toxic nitrites and thence into relatively harmless nitrates. In denitrification, anaerobic bacteria convert nitrates into free nitrogen via nitrous oxide. These processes often occur simultaneously within different zones of the same filter or within the same filter medium (such as sintered glass granules, cylinders or hoops).

● *How do I know when zeolite needs recharging, and how do I go about it?*

An appropriate water testing kit will show increasing levels of ammonia in the pond water as the zeolite fails to adsorb it. This applies primarily to new filters with immature bacteria populations, and to ponds whose filters are working at full capacity (e.g. small filters in heavily stocked ponds). Zeolite can be recharged by immersing it in a solution consisting of about 30g salt (sodium chloride) dissolved in about 5 litres of water. Leave for a day, during which time the zeolite releases ammonia and takes in sodium. Rinse it thoroughly before reusing.

● *Can activated carbon be recharged?*

Yes, through prolonged heating in an oven, although activated carbon never fully regains all its adsorbent properties. It is therefore best to discard spent carbon and replace it with a fresh batch.

Biological filtration involves the conversion of toxic ammonia – a waste product of fish – into nitrites (also toxic) and then into nitrates (relatively harmless, except at high concentrations). The whole process is carried out by bacteria. The two genera involved were traditionally thought to be *Nitrosomonas* and *Nitrobacter* respectively, though recent research has indicated that these are either not the principal ones, or may even not be involved at all. Whatever the truth of the matter, there is no doubt that bacteria of some type are the main agents of detoxification. Biological media include matting, ribbons, open-pore foam, hair rollers, plastic pipe-type media, Canterbury spar, gravel, ceramic granules and cylinders, lava stone chips, sintered glass hoops or solid cylinders. By virtue of their sheer physical presence in the tank, biological (and chemical) media can also act as mechanical media. Indeed, if they are allowed to become coated in debris, their effectiveness is impaired. They will not break down solid waste particles and need a good flow of well-oxygenated water to perform at their best.

As with mechanical media, biological media should be rinsed regularly with pond, tap or rainwater. Similarly, biological media should never be cleaned with detergents, however mild they might be in human terms.

BIOLOGICAL FILTERING MEDIA

Plastic bio-media

Sintered glass media

Rock composite

▲ ▶ *There is a wide variety of pond fountains on the market. Shown on this page (clockwise from top): a pirouette fountain with rotating jets; a bell fountain; a simple spray jet; a tulip or inverted cone jet; a variable calyx jet; and a three-stage fountain. On the facing page, a fountain has been combined with an unusual sculpture of three frogs to produce a focal point for a garden.*

Lighting and Other Accessories

Some accessories designed for the pond are functional, while others are purely aesthetic. Whatever their role, they are all meant to increase the pleasure of the hobby, either by making life easier for the pond keeper, or by adding colour and interest to the pond and its surroundings.

One especially striking way of enhancing the beauty of certain ponds is to install illumination. Through the use of different types of lighting, above and below the water, a pond can take on a whole new aspect at night, with the water reflecting the lights and pondside plants casting unusual shadows. Of course, context is important here – it would be wholly inappropriate and counter-productive, for example, to attempt to light a wildlife pond. But in the right circumstances and applied with subtlety, pond lighting can be a real asset.

Types of lighting for the pond include submerged models designed to illuminate the water itself, and floating lamps that lie on the surface. These disperse most of their light above the water, but also create small pools of light below. Often, but not invariably, both types of system offer a choice between coloured and clear lamps.

Increasing interest in renewable, "clean" energy sources is also reflected in an expanding range of solar-powered lights from specialist suppliers. Most such lamps do not have high-illumination outputs, but this is not usually a crucial factor, since solar lamps are most commonly used as subtle "accent" lights.

An enormous selection of less functional but aesthetically pleasing accessories is available to the pond keeper. These range from simple Japanese pool adornments such as "deer scarers" (*shishi odoshi*; see Small Water Features, page 21) and washing troughs or basins (*tsukubai*) to highly elaborate statues, lanterns, and other features made from a variety of materials.

Finally, though not strictly speaking an accessory, every pond keeper should keep at hand the telephone numbers of individuals or companies that can help in cases of emergency.

Some Useful Accessories

Particularly useful and important accessories that no pond keeper should be without include the following:

- ✓ A selection of pond nets – including at least one long-handled one.
- ✓ Inspection bowls.
- ✓ A comprehensive medicine chest.
- ✓ Water testing kits – at least for ammonia, nitrites and nitrates.
- ✓ A pond thermometer.
- ✓ A pond "vacuum" for removing debris.
- ✓ A pond heater – in zones that experience freezing conditions during winter.
- ✓ A butyl liner repair kit.
- ✓ A net to stretch over the pond to protect it against leaf fall.
- ✓ Spare filter valves, washers, connectors, etc., plus a set of tools.
- ✓ Planting baskets and associated materials.

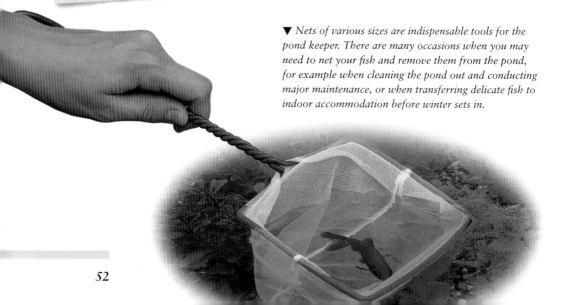

▼ Nets of various sizes are indispensable tools for the pond keeper. There are many occasions when you may need to net your fish and remove them from the pond, for example when cleaning the pond out and conducting major maintenance, or when transferring delicate fish to indoor accommodation before winter sets in.

● *Are all accessories safe to use?*

The great majority are perfectly safe and often carry official accreditation to this effect. However, where metal objects are concerned, or items containing some metal or other components that may be toxic to fish and other pond inhabitants, you should ask and seek assurance before you buy. Some potentially toxic compounds can be rendered safe by applying a coat of appropriate sealant.

● *What is the distinction between an accessory and a piece of essential equipment?*

The dividing line between what constitutes an essential item of pond equipment and an accessory is not always easy to define. One reason for this is that the same item can be essential under one set of circumstances but not under another. Thus, in heavily stocked ponds, filters are indispensable. However, in sparsely populated systems, they can be seen as accessories, but only if all other factors are favourable. Generally speaking, anything that helps the pond keeper, but is not absolutely vital for the survival of the pond inhabitants, is definable as an accessory.

▲ *Lighting can have a dramatic effect, whether it is used to cast an enchanting glow across an entire pond or (inset) to highlight a particular feature, such as this bell fountain.*

▶ *A light cluster for installing around a pond fountain. Accessories such as this can be fitted with interchangeable coloured lenses, which change the "mood" of the pond lighting.*

Pond Fish

ONLY A TINY PROPORTION OF THE
20,000-plus fish species known to science are suitable
for keeping in ponds. Nevertheless, these few species
still cover a wide range of shapes, sizes, colours and
habits. The choice of fish available to the pond keeper
is therefore extensive, and expanding all the time.

However, as fish keeping in water gardens grows, it is
vitally important to realize how greatly pond fish can vary in
their requirements. So diverse are they that a pond which is
perfectly adequate for one species may be little more than a torture
chamber for another. Even a type of pond that will safely house a
particular species in one part of the world can prove a watery
grave for the same species in another region.

Thus, Japanese Medakas (*Oryzias latipes*) will live perfectly
happily in a well-kept shallow, small pond in tropical and subtropical
regions, while Koi (*Cyprinus carpio*) will almost certainly die under
such conditions. Equally, White Cloud Mountain Minnows
(*Tanichthys albonubes*) will thrive in a shallow pond in a frost- and
snow-free region. However, transfer this same pond to an area that is
prone to freezing temperatures and, despite their innate toughness,
few, if any, White Clouds will survive their first winter.

The following section helps you to match fish successfully to
the pond environment you have created. Facts and practical tips
are given on buying fish, introducing them to your pond,
breeding them and caring for their health. Profiles of a number
of species then outline their characteristics and the
appropriate conditions for keeping them.

▶ *A dazzling collection of Koi varieties. These fish
have grown hugely in popularity in recent years.*

Choosing and Introducing New Fish

ONCE A POND IS UP AND RUNNING, THE URGE to rush out and buy some fish seems irresistible. However, buying livestock with no forethought can result in a high mortality rate. Fish are not conventional "pets", but this does not mean that owners have any less responsibility for their well-being. Care does not just mean providing food and breeding partners; it also involves maintaining the fishes' health and protecting them from predators (including other fish). Taking a broader view, you should also be alive to their potential effect on local fauna and flora.

Taking Stock

Every year, a host of new pond keepers and water gardeners swell the fast-growing ranks of "outdoor aquarists". Pond fish will be at the top of their shopping lists. In addition, there is a constant demand for livestock among more experienced hobbyists, who may either want to expand their collection by adding new specimens, or need to replace fish that have succumbed to old age or the rigours of winter.

Either way, it is vitally important to assess your particular requirements accurately and buy wisely. For instance, it is essential to determine how many fish (and of what size) a pond will comfortably accommodate before acquiring any. As a general rule, you should allow 24sq in (155sq cm) of available surface area per 1in (about 2.5cm) of fish (excluding tail).

It is unwise to begin too close to this upper limit; rather, you should allow some room for the fish to grow. In theory, stocking calculations should be based strictly on the eventual size that the fish will attain, but in practice this can result in a rather empty-looking pond. It is a good idea to purchase a mixture of fish, comprising some small specimens that will still grow and some full-grown larger ones. You must ensure that all the fish will be compatible.

Under certain circumstances, you may exceed the recommended stocking level. For example, an effective filtration system, the use of an aerating device (venturi), or just a deep, extensively planted pond are all factors that permit higher density. Nevertheless, you must stock your pond in gradual stages, allowing the fish and filtration system to adapt to each increase and only proceeding on to the next stage when you are certain that the pond can deal with the additional levels of waste products.

How to Identify Healthy Fish

The first point to emphasize is that no one can give a cast-iron guarantee that any fish, however healthy it may look, is absolutely free of disease or parasites. Since every living thing plays host to a natural, safe complement of parasites, it could be argued that if a fish is wholly free of parasites,

▲ Floating inspection bowls such as this allow fish (usually Koi) to be examined at close quarters by potential keepers before they make a purchase.

Choosing Healthy Fish

- **Lively Disposition** Healthy Goldfish, Koi, Orfe and similar fish are constantly on the move and exhibit quick reactions. If such fish "hang" in midwater, float on the surface, or lie on the bottom, with little visible movement, they are best avoided. In the case of a bottom-dweller, like a Tench, there is no cause for alarm – it will just be resting.

- **Erect Fins** A lively fish will tend to carry its fins well extended. The dorsal (back) fin is often flicked open and shut, usually as the fish hovers or changes direction, and is generally held down as a fish flicks the caudal (tail) fin to propel it over short distances. However, in a long-tailed variety, the caudal fin will not be carried as fully extended.

- **Bright Eyes** Healthy eyes are clear, while injured or unhealthy ones tend to be "milky" in appearance. However, this is not always easy to spot, particularly from above, or where small and/or bottom-hugging fish are concerned. Despite this, it is wise to take some time to look for signs of cloudiness.

- **Bright Colours** This can be difficult for the untrained/inexperienced buyer, since the scales of all fish carry a protective layer of mucus. "Bright" means the absence of a milky or cloudy covering of excess mucus on the body. Excess mucus creates a dull overall effect. Once seen, it is unmistakable.

- **Balanced Swimming** Healthy fish do not normally "roll about" or lose their balance when swimming; neither do they float or sink. However, immediately after a meal, healthy fish will often hover and can sometimes rock gently as they chew their food.

- **Full Body** This does not mean a full-looking belly, but refers to the flesh or muscle tone of the fish. A healthy fish should appear "solid". With the possible exception of large-headed fancy Goldfish varieties, such as Orandas, healthy fish do not have oversized heads. In particular, the area immediately behind the head should not look pinched. In other words, the "neck" area of the fish should be either about as wide as or (in some species and varieties) even slightly wider than the head itself.

▲ *A Common Goldfish* (Carassius auratus *var.*) *caught and bagged for transportation. With its erect fins, unblemished body and bright eyes, this specimen would be a good choice.*

- **Good Appetite** Without a doubt, healthy fish have healthy appetites. Consequently, buyers are sometimes advised to ask to observe the fish they want to buy while the fish are feeding. This advice may sound quite sensible, but it is not. For a start, if the fish have already been fed, they may not be as voraciously hungry as one might wish them to be. This does not mean they are not healthy; they are just not hungry. But suppose they are hungry and devour the food offered; and having demonstrated their good health, the chosen specimens are netted and bagged up for the journey home. Quite naturally, the stress of being netted, especially immediately after a meal, soon results in the fish defecating and excreting in the bag water during the journey home. Such travelling conditions are hardly ideal for the fish, particularly if the journey is long. Perhaps, therefore, it is wiser to forget about asking for a demonstration of a healthy appetite and concentrate on the other factors instead.

- **Other Signs** Damaged fins and missing scales are unsightly, but not necessarily indicators of poor health. Rips in the finnage and loss of scales can easily occur at any time, as a result of fighting or injuries sustained on pieces of pond decor. They usually regenerate without any infection setting in. However, if there are any signs of blood spots, streaks, or fungal ("cottonwool") growth at the place where the scales are missing or the fins torn, then you may be fairly certain that the fish is in poor health and should be rejected for purchase.

How to Avoid Unhealthy Fish

- Raised scales (dropsy).
- Swollen abdomen possibly caused by fluid retention (if this is not a feature of the species or variety, e.g. round-bodied types of Fancy Goldfish).
- Protruding eyes (pop-eye or exophthalmia), as long as these are not a characteristic of the fish, as in Telescope-eye Moors.
- Ulcers (bacterial lesions).
- Overlarge head on undersized body, a possible sign of what is known as "consumption"; note that some Goldfish varieties, such as Orandas, Lionheads and Ranchu, normally have large heads.
- Lack of movement or "skulking" away from the shoal.
- Obviously thin, bony body with "pinched" area just behind the head.

- Listless swimming and/or loss of balance.
- Physical damage to fins, scales and/or body.
- Fungal ("cottonwool") growths, blood spots and/or streaks.
- Persistent scratching against objects in the water (rocks, etc.).

▶ *This Red Fantail Goldfish* (Carassius auratus *var.*) *is showing signs of distress caused by a swimbladder problem.*

then it must be dead. Pathogenic (i.e. disease-causing) organisms do not normally give rise to epidemics in healthy hosts, because the micro-organisms cannot gain a firm enough foothold to multiply in sufficient numbers to cause disease.

There are, however, a few danger signs that you should look for when buying fish. For example, if some fish in a holding trough or tank look decidedly unhealthy, it is best to avoid buying not only those but also any of the other fish from the same tank, even if some of the others appear healthy. The accompanying panels on this and the preceding page give detailed information on how to identify healthy and unhealthy fish. All reputable dealers should be more than willing to advise you in making your selection. Experienced fishkeepers and members of local aquatic societies are also usually very happy to help newcomers choose their fish.

Protecting Local Fauna

The widespread introduction of certain fish, such as the Common/Mirror/Leather Carp (*Cyprinus carpio*), the Rainbow Trout (*Salmo gairdneri/ Oncorhynchus mykiss*), the Zander/Pikeperch (*Stizostedion lucioperca*) or even the Common Goldfish outside their natural ranges has had a far-reaching (and sometimes adverse) effect on original native species. One of your responsibilities in pond keeping and fish husbandry is therefore to do everything you possibly can to prevent your fish from escaping into surrounding waters, especially if those fish are non-native.

In several parts of the world, legislation has been enacted to prevent the sale or keeping of certain species or varieties of fish (see Wildlife and the Law, pages 178–179). Although such regulations may sometimes appear excessively stringent, they are principally devised to protect indigenous fauna. Legislative controls may have local, regional, national or international force, and no hobbyist could possibly be expected to know the details of every law on wildlife protection worldwide. Nevertheless, responsible pond keepers should still make an effort to acquaint themselves with the general provisions of international legislation. Even more important is to familiarize oneself with the regulations governing the pond hobby in one's own country or region. Local laws can change quite rapidly, so if you have any doubts about the purchase of a particular species or variety of fish, ask your dealer first, or check with the relevant local authorities before proceeding.

In pond keeping, as in many other hobbies, environmental awareness is taking on an increasingly higher profile. For many years now, experts from the aquatic retail industry have been invited to take part in the consultation process that precedes the drafting of important legislation on the import and sale of ornamental fish. As a result, the fish that you will find offered for sale in stores in most countries have all been either legally imported or bred specifically for the hobby.

● *What is a coldwater fish?*

The term "coldwater fish" originated in temperate countries to refer to species and varieties that could be safely kept in aquaria without supplementary heating. This distinguishes coldwater fish from their tropical counterparts. Because pond fish in temperate areas do not require any supplementary heat, the term "coldwater" is commonly used when referring to them. However, in the case of tropical and subtropical regions, this term is totally inappropriate, since there are no such things as coldwater ponds in the tropics, where providing a cool pond environment would require the installation of special cooling units – at considerable cost to the keeper.

● *Can large and small fish be kept together?*

This depends to a large extent on the species in question. Predators such as pikes (*Esox* spp.) will devour all they can, and so should only be kept with fish that are far too large to swallow, while Koi and Goldfish can be kept in shoals consisting of fish of widely differing sizes. However, if they are large enough, even docile fish such as these will consume tiny fry which they probably cannot identify as young of their own species, but simply as food.

● *Can pond fish see in colour?*

Yes. Many pond fish – particularly the cyprinids, which include Goldfish, Koi, Tench, Rudd, Roach, Minnows, Shiners and others – are known to possess surprisingly sophisticated colour vision. This is because their retina (the light-sensitive membrane at the back of the eyeball) is generously equipped with cone cells, which are responsible for the ability to perceive colours.

● *Do all fish hibernate?*

No. The only ones that do so are those kept in unheated ponds and exposed to freezing temperatures for an extended period of time. Even in such areas, the trend towards installing pond heating units means that many fish go through even long winters without hibernating at all. During cold snaps, a fish's metabolism automatically slows down, but it is debatable whether such temporary lethargy (as happens during a short-lived period of low temperatures) can be regarded as an example of true hibernation.

● *What is an exotic species?*

Strictly speaking, an exotic species is one that originates in a foreign country. According to this definition, therefore, the Goldfish is "exotic" in all areas of the world except China and parts of Siberia, where it originated. However, few people today would consider this extremely common fish to be "exotic" in the sense of being rare, mysterious and intriguing – which is how the term has come to be used. The opposite of "exotic" is "native". Whether a species is exotic or native depends on your point of reference.

● *What effects do exotic species fish have on native ones?*

The effects can range from negligible to devastating. If an exotic species cannot survive or breed where it is released, any effects will only be short-lived. On the other hand, if both survival and reproduction are possible, then long-term competition for food, space, spawning sites, etc., all affect the outcome. If an exotic species is more vigorous than native fauna, it will reduce the size of the latter's population or even completely displace it. If they are more evenly matched, a new balance is established.

Transporting and Releasing Fish

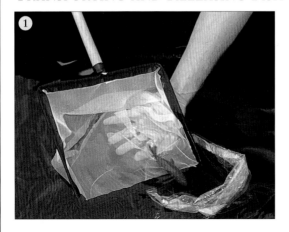

Netting a fish and transporting it between the shop (or a friend's pond) and your home inevitably causes it stress. However, if you carefully follow this series of steps, you will be able to minimize the shock and introduce the fish safely to its new home.

1 Gently net the fish, without chasing it, and transfer it to a heavy-duty polythene transportation bag. The bag should be filled with one-quarter to one-third water, and the rest air or, better still, oxygen. (Larger fish should not be netted, but guided gently into a floating bowl, from where they can be coaxed directly into a bag.)

2 The aquatic store assistant will inflate the bag with oxygen for you. This sustains the fish and lends the bag rigidity.

3 To further cushion the fish from shocks, place the bag in a sturdy cardboard box. On the way home, take care not to subject it to violent jolts or knocks, or to abruptly expose it to bright lights; both will frighten the fish.

4 On arrival, carefully lift the bagged fish from the box and float the bag – still fastened – in the pond for at least 10 minutes. This allows the water in the bag to reach the same temperature as the pond water. Avoid direct sunlight.

Once these first steps have been completed, the release procedure may begin:

5 Untie the bag, without releasing the fish; on no account should you ever pop the bag open. Allow enough pond water into the bag to increase the volume by one-quarter. Then, to let the fish become accustomed to the new

water, drape the bag over the side of the pond and hold it in place with a large stone. Leave for 10 minutes or more, then add another small amount of pond water to the bag.

6 Finally, tilting the bag slowly, let the fish swim out into the pond. *Never* pour the fish out of the bag. The fish will find shelter or join an existing shoal. You can mix the water remaining in the transportation bag into the pond, but if you are concerned about introducing disease, simply discard it.

Wait at least several hours (overnight is preferable) before feeding your new fish. Offer it small amounts at first, and gradually increase the portions if the fish seems hungry.

Note that some delicate species need longer to acclimatize. Seek advice from the aquatic store.

● *Will mixing transportation bag water with pond water pose any health risks for the existing pond fish population?*

It could, but it generally does not. Bag water does contain some potential pathogens, but no pond can be 100% free of pathogens either. However, as long as the fish are robust and healthy, and as long as the volume of new water is small in proportion to that of the whole pond, the risks are minimal, especially if the pond is equipped with an efficient water treatment system.

● *Would netting or manually lifting new fish from a bag into a pond reduce the risk of introducing disease?*

Two points should be noted here. First, nets and hands (to say nothing of fish!) that come into contact with bag water will themselves transfer some pathogens into a pond. Also, netting or handling fish between bag and pond is a stressful experience for the fish. It can result in the removal of scales and/or mucus, which constitutes a potential health risk in stressed fish. While no transfer method is totally risk-free, the technique described opposite is generally regarded as being preferable to all others.

● *Why is it advisable to introduce new fish into a pond in a series of small steps, rather than one single major one?*

Being moved to a new environment is extremely traumatic for fish, and as stress is one of the principal causes of fish death, anything that can be done to reduce the effect is well worth the effort. Although new fish have been released into ponds in one step with no ill effects – by simply opening the transportation bag and allowing them to swim out – this practice is not recommended. If you implement a series of steps involving light and temperature equilibration, along with a gradual mixing of bag and pond water, this allows the new fish to become acclimatized to their new environment with a minimum of stress and to adjust their metabolism accordingly.

● *What is the problem with netting large fish?*

For any fish, the loss of the buoyancy that is provided by water creates unnatural pressure on its body tissues and internal organs, which can result in injury. This is especially true of large fish, which tend to be heavier in the body.

Foods and Feeding

THE OLD ADAGE "WE ARE WHAT WE EAT" IS every bit as true of fish as it is of humans. Of course, food is not the only determinant; our genetic make-up and a whole host of other factors also contribute; but an adequate diet is crucial for long-term health and survival. For the fishkeeper, some basic knowledge about nutrition is therefore needed in order to understand the dietary requirements of the fish we keep.

Among pond fish, there are numerous feeding habits, from the almost continuous, predominantly herbivorous/detritus-scouring activities of many carps, to the more carnivorous techniques adopted by predominantly predatory species like the Orfe (*Leuciscus idus*).

Nutritional Requirements

Whatever their feeding habits or diets, all fish require the same basic nutrients: proteins, carbohydrates, fats or oils (lipids), vitamins and mineral/inorganic salts (trace elements).

Proteins are large organic molecules required by all living organisms. The main reason for this universal need is that many proteins actually form part of the living structures found in and around cells; and, since growth consists of the production of new cells, a lack of protein will result in a lack of growth. All living processes also require energy, but the release of this energy is impossible in living tissues without the involvement of a group of substances known as enzymes, which are also proteins.

One common element that all proteins share is nitrogen. Although very common (it makes up 78% of the Earth's atmosphere), it is unreactive, and so it cannot be assimilated directly from the atmosphere by most animals and plants. Fortunately, some species of bacteria and other organisms can assimilate (fix) nitrogen directly from the atmosphere. They form part of a complex series of interactions collectively known as the nitrogen cycle (see diagram).

Nitrogen fixation involves the incorporation of this element into nitrites and, later, nitrates. Some of these nitrates are taken in by plants and assimilated into their body tissues in a number of forms, such as structural proteins. These plants may then be eaten by herbivores, which might include certain species of fish. Some of the plant-eaters may be eaten by other fish, which thus obtain their nitrogen indirectly.

Herbivorous fish possess long alimentary canals, while carnivorous ones have correspondingly shorter guts. Therefore, it does not make good sense to choose a vegetable-based formulation and feed it to all types of fish all of the time.

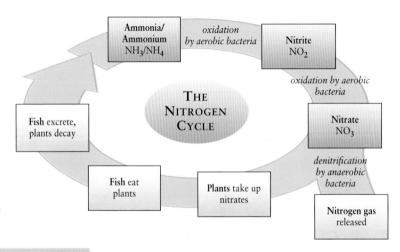

◀ *The nitrogen cycle describes the sequence of chemical reactions by which nitrogen circulates through an ecosystem. In the pond, the target is to remove harmful ammonia and nitrites (plus less harmful nitrates) from the system, thus preventing a dangerous build-up. The amount of protein in the diet of fish affects the amount of nitrogen passed in their waste.*

... It depends on the species and the temperature. As temperature rises, so do metabolic rates, while levels of dissolved oxygen drop. For many temperate fish, "metabolic overload" sets in above 30°C (86°F), causing a loss of appetite.

● *Will all pond fish eat the same food?*

Modern high-quality commercial foods are palatable enough to be accepted by most fish. Many floating Koi pellets and sticks, for example, will be eaten just as readily by such diverse fish as Rudd (*Scardinius erythrophthalmus*) and Guppies (*Poecilia reticulata*). However, the dietary requirements of every species should be thoroughly investigated, and specialized foods or supplements used as needed. To a certain extent, the natural fauna and flora of an established pond will also help overcome minor nutrition deficiencies that may have been overlooked.

▲ *Goldfish* (Carassius auratus *var.*) *feeding on flakes near the surface. Like most coldwater fish, Goldfish are not fussy eaters and will take meat or plant food.*

If this is done, the fish are likely to develop dietary problems, reducing their resistance to all sorts of pathogenic organisms they encounter. It is also pointless to supply adult fish with the same high protein levels recommended for the fast-growing young. If this is done, some of the excess will be expelled as nitrogen-rich toxic waste products such as ammonia. Growing fish customarily require a diet that contains around 35% protein, while fully grown fish need no more than 30%.

Equally dangerous is a diet that is too low in protein. Over the longer term this will result in growth retardation, decreased ability to repair tissues and increased susceptibility to infection.

Carbohydrates, which are one of the two principal sources of energy in fish, are mainly (but not exclusively) sugars and starches. All carbohydrates consist of just three elements: carbon (represented in chemical shorthand by the symbol C), hydrogen (H) and oxygen (O).

Like humans, all fish need carbohydrates but cannot synthesize them from their constituent parts, so carbohydrates must come from their diet. The key carbohydrate-rich food for fish is green plants, which, unlike animals, can synthesize carbohydrates through photosynthesis (see Why Are Plants Important?, pages 118–119).

Herbivorous fish obtain their carbohydrates directly by eating green plants. In the case of omnivores, only part of their carbohydrate requirement comes from eating plants; the remainder is supplied by aquatic insects and other small animals which have fed on plants or plant-eating organisms. Similarly, the diet of piscivores and carnivores is based on other fish (or other animals) that contain carbohydrates derived directly or indirectly from plants.

Excess carbohydrates are stored by fish in the form of a chemical called glycogen. However, there is a limit to how much glycogen a fish can store. Once this limit is exceeded, extra carbohydrates will be converted to fats.

Fats (Oils/Lipids) are the other main source of energy in fish. In addition to this role as an energy source, fats also form essential components of cell membranes and therefore play a part in the formation of new tissues. As you might expect, therefore, if their diet is well balanced between fats and other nutrients, fish can utilize a higher percentage of proteins in growth or in building and repairing tissues. If the diet is lacking in fats and carbohydrates, some proteins will be used up as alternative energy sources instead of being used for growth or repair.

Fats can also be stored for use during periods of food scarcity. However, if too much fat is present in a fish's diet, the excess will be deposited around organs such as the liver, kidneys and heart. This has an adverse effect on the fish's health and can lead to serious disorders, even to

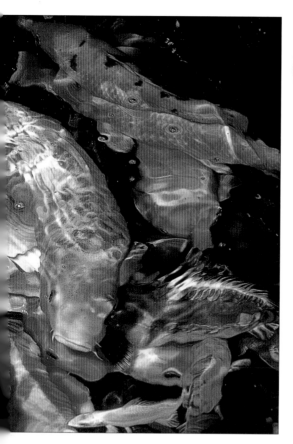

▲ *Koi* (Cyprinus carpio) *quickly become so tame that they can be fed by hand. Once this happens, they will rush over as soon as you put your hand in the water.*

death in extreme cases. At the other end of the scale, lack of essential fat-derived food components (fatty acids) has been linked with diseases such as Fin Rot.

Because fish are less active in ponds than in the wild, they are more prone to developing fatty deposits if their diet is not balanced. Common Carp (*Cyprinus carpio*) kept in captivity have been found to carry between 10 and 45 times more abdominal fat than wild-caught specimens. This not only affects the liver, kidneys and heart, interfering with vital bodily processes; it also causes decreased resistance to infection. In addition, excess fat has a detrimental effect on the gonads (sexual organs) and the production of gametes (sperm and eggs). Over-fat fish are therefore next to useless for breeding. Excessive fat infiltration of the liver may develop to such an extent that all the "filtering and balancing"

● *What does Q_{10} stand for?*

The Q_{10} factor of a reaction or process is the rate at which it changes over a 10°C (18°F) temperature rise or drop. For reactions involving enzymes, as in food digestion, the Q_{10} value varies between 2 and 3. This means that, for every 10°C (18°F) rise, the reaction rate doubles or trebles. The opposite applies for drops in temperature. This explains why pond fish in temperate regions require very little or no food during winter, progressively higher quantities as temperatures rise in the spring and summer, and progressively less during autumn. (See Pond Management, pages 186–199 for further details on seasonal feeding.)

● *How can pond fish be encouraged to feed during excessively hot weather?*

Feed them during the coolest times of day – early morning and late evening. Evening is better, because the fish then have the entire night to digest their food at more manageable temperatures. Aeration and water turbulence by means of cascades, waterfalls, etc., will also help by increasing the concentration of dissolved oxygen to meet the fishes' high metabolic demands in hot weather.

● *If many fish feed on an almost continuous basis in the wild, why should pond fish be fed only at intervals?*

Naturally-occurring foods – even nutritious ones – contain large amounts of water (up to 90% by weight). In contrast, commercial foods like flakes, pellets, granules and sticks contain very little water (sometimes as little as 5%). This makes their nutrient-to-weight ratio much higher than that of naturally-occurring foods. As a result, a small amount of good-quality commercial food has the same nutritional value as a much higher weight of natural food, particularly if the latter is of plant origin.

● *Can I feed my fish by hand?*

With large fish such as Koi, you can certainly attempt hand feeding. Teach them to associate you with food by feeding them several times a day, remaining by the pond. Do not make any sudden movements, which can frighten them. Gradually come closer and then dip an empty hand in the water nearby while they feed. Finally, begin to offer small amounts of pre-soaked pellets in an open hand, a few inches below the surface (dry pellets will float off). Once the first fish has taken food, others will follow.

properties of this important organ may be lost. Fish affected in this way invariably die.

Fat-rich and fat-generating diets must therefore be very carefully monitored if their harmful effects are to be avoided. Young fish, fortunately, are not prone to fat-related ailments, as they use up large amounts of energy in growth.

Vitamins are vital to fish health. Even if their diet is perfectly balanced in terms of proteins, fats and carbohydrates, serious health problems will still arise if one or more vitamins are either present in insufficient quantity or lacking altogether. For example, a diet that does not contain enough Vitamin A can lead to defective vision, while one that is too low in Vitamin D will affect bone formation through the disruption of the processes involved in calcium and phosphorus assimilation. A lack of Vitamin B_6 (pyridoxine) impairs the metabolism of protein as well as fat; if there is not enough Vitamin B_2 (riboflavin), carbohydrate metabolism is similarly affected. A lack of Vitamin C (ascorbic acid) has been shown to cause spine deformities (Scoliosis) in certain species. Fortunately, adequate supplies of all vitamins are provided in good-quality commercial foods for fish.

Mineral or Inorganic Salts are also referred to as "trace elements" because they are required in only very small amounts. Although the function of the 25 or so minerals is known, individual amounts are very difficult to estimate because aquatic organisms absorb minerals both from the food they consume and from the water itself.

A lack of calcium and/or phosphorus will inhibit bone formation (among other metabolic processes). At best, fish that are suffering from such a deficiency will exhibit retarded growth. In extreme cases, so

Types of Fish Food

Fish foods come in a variety of forms: dry formulations, livefoods, freeze-dried and deep-frozen preparations are the main types. Some are intended for general use, others for specific purposes.

▲ *Tubifex* (x200%)

◄ *Daphnia* (x700%)

◄ *Bloodworm* (x200%)

Dry Formulations are available as flakes, pellets/ granules and sticks. All come in a variety of sizes, from fine, almost-powdery foods designed for very young fish, to large grains or sticks for large adults such as Koi. There are "staple" recipes for everyday use; high-protein diets for growing fish and general feeding during late spring and summer; carbohydrate-rich (easily digestible), slow-sinking formulae for colder weather use; Spirulina-enriched foods for enhanced coloration; and so on. All the top manufacturers are continuously researching and improving their products to provide pond fish with a sound diet throughout the year. A good-quality dry diet is complete – there is no need to supplement it with other types.

Livefoods are commonly used as a natural dietary supplement. Despite the versatility and good value of the best dry formulations, most pond keepers like to vary the diet for their fish by providing livefoods, at least occasionally. While it is debatable whether this has any measurable effect on nutrition, fish appear to relish livefoods.

The most widely available types of livefood include *Daphnia* (waterfleas), adult *Artemia* (brine shrimp), *Tubifex* (tubificid worms) and *Chironomus* (bloodworms), although there are others. If these come from cultured stocks, they are likely (with the possible exception of *Tubifex*) to be pathogen-, parasite- and predator-free. This does not, however, apply to stocks collected from the wild (except *Artemia*, which is marine and is therefore cultured in special tanks). This is worth keeping in mind before you offer wild livefoods to fish in a healthy pond environment.

Freeze-dried Foods consist of livefoods whose moisture content has been virtually eliminated. The processing of these foods also results in the destruction of most pathogenic agents, thus making them very safe, while retaining their nutritional value. Freeze-dried foods come either loose or

many enzymatic processes may become affected that the fish can die. Iodine is necessary for balanced hormonal activity, while iron is essential for oxygen transport, energy transfer and the formation of haemoglobin (the red blood pigment) itself. Even chlorine, which is highly toxic when present in excess, is essential, along with sodium and potassium. These minerals control osmoregulation (the maintenance of an adequate salt and water equilibrium in the body tissues).

If a pond is maintained efficiently, with adequate plant and fish stocks, an appropriate feeding regime and a suitable water management system, the chances are that there will always be sufficient trace elements present to meet the needs of both fish and plants. If you suspect a deficiency, you may add supplements to the fishes' food, but these should not be relied on long-term as a substitute for a properly balanced diet.

as small compressed blocks. Loose, small organisms, such as *Daphnia*, *Chironomus* and *Tubifex*, are particularly suitable for small fish. Larger items, such as shrimps of various types, can be fed to larger fish, which also like cubed preparations.

Deep-frozen Foods originally consisted of single (i.e. unmixed) items such as bloodworm, *Mysis* shrimps and the like. In more recent years, mixed diets have been introduced, partly as a result of research carried out on formulations for marine fish. Most deep-frozen foods are gamma-irradiated, which destroys pathogenic organisms and makes the food very safe to use.

Other Foods include fresh vegetables, such as garden peas and bruised lettuce, along with crumbled biscuits, brown bread and other unprocessed items. Modern innovations include high-nutrition foods, neither deep-frozen nor freeze-dried, packed in sterilized fluids in jars; and foods containing immuno-stimulants to increase resistance to infection.

Q&A

● *How do I calculate the correct amount of food per feeding?*

... Few of us are expert or patient enough to work out requirements involving the weight and age of a fish, the type of food, the water temperature and other factors. Fortunately, fish don't eat if they are not hungry. Five minutes' continuous feeding on a good-quality commercial diet gives a fish enough nutrients to keep it going for at least a few hours. If any food is left over, it is safe to assume that the fish has had enough.

● *Is it essential to feed pond fish?*

Not necessarily. A large, well-planted pond contains a good stock of natural plant and animal food items, supplemented by aerial insects and wind-borne plant debris. If the fish are few, they may be able to get by without additional food, though it is debatable whether they will thrive on a long-term basis. If the pond is well-stocked, regular feeding is absolutely essential.

● *How should fish be fed during family vacations?*

If it is a matter of only 2–3 weeks, and the fish have been well fed previously, no special arrangements are needed. Well-fed fish build up fat reserves which, added to the natural animal and plant foods (including algae) in ponds, will see them through several weeks without any hardship. For longer breaks (or even for short ones, if desired), automatic feeders and "vacation blocks" may be used. Alternatively, leave clearly marked, individually packaged, small portions of food, accompanied by strict instructions, with a friend or neighbour. If the "pond minders" are not experienced in fish feeding, it is worth reducing the amount of each packaged portion of food by half and doubling the interval between feeds, just to be on the safe side. Overfeeding must be avoided at all costs, therefore always err on the side of caution.

▲ *Some common dry foods: 1 Large Koi pellets, 2 Koi sticks, 3 Freeze-dried brine shrimp blocks, 4 Fish flakes, 5 Mini-pellets.*

Breeding Fish

BIOLOGICAL CLOCKS (OR "BIORHYTHMS") ARE remarkable things. They control behaviour and metabolic activities as varied as fluctuating hormonal levels during a 24-hour period, or large-scale annual migrations. Daily cycles (circadian rhythms) help animals and plants survive from day to day, while longer, seasonal/annual cycles help individuals and species survive from year to year and from generation to generation. It is these second, "longer-set" biological clocks that are clearly apparent every spring, resulting in the characteristically hectic activity evident among our pond fish, mainly – though not exclusively – during March, April and May in the northern hemisphere, and from September to November in southern zones, the peak of the breeding season. But why spring and not autumn? And what actually triggers spawning activity? While these are simple enough questions, the answers are extremely complex and, as yet, incomplete.

Triggers of Spawning

From the fishes' perspective, spring is the ideal time for spawning. It is a time when ponds in temperate zones are awakening from hibernation (or, at least, from the coldest months of the year). To some water gardeners, this signals the beginning of a difficult period, characterized by green water and blanket weed. But, looking at it from the fishes' point of view, it is this very same flush of new life among the algae – and among the microscopic creatures that feed on them and are, in turn, fed upon – that makes this season so good for breeding. This rich algal "soup" of micro-organisms forms the basis of a newly-hatched fish's diet, and continues to ensure it a constant supply of progressively larger food items during its first few months of life.

As water gardeners who like to see crystal-clear water in our ponds, we may wish to disrupt this natural process by eradicating the unsightly algae that disfigure our creations. Yet we should always be mindful of the biological importance

Q&A...

● *Is there a risk of mistaking breeding tubercles for whitespot disease?*

Once the differences between the two are known, there is no risk at all. However, inexperienced fishkeepers who have not come across either before often mistake tubercles for the protozoan parasitic disease known as whitespot or Ich (Ichthyophthiriasis) and dose the pond with medication in order to get rid of the "problem". When it persists (as it invariably does, of course), they dose again, causing tremendous and unnecessary stress among perfectly healthy fish. Even without a scale to measure the size of the spots, it is easy to tell tubercles and parasites apart, since the (larger) tubercles are generally confined to specific areas (around the gill covers and the first ray of the pectoral (chest) fin in many cyprinids) and do not have the overall appearance of a randomly distributed "dusting" of small spots that is characteristic of whitespot.

● *What are spawning mats and brushes?*

These are flat pieces of toxin-free material commercially produced to aid spawning; the mats are equipped with a layer of soft fibres, and the brushes with a mass of bottle-brush-like protrusions. They come in a variety of sizes to suit a range of species; where pond fish are concerned, they are usually used with Goldfish and Koi . One great advantage over natural fine-leaved vegetation is that they can easily be removed from the pond after spawning, thus allowing the eggs to hatch in safety away from the attentions of hungry parents and other pond inhabitants. Another is that, once used, mats and brushes can be sterilized and re-used immediately, or stored for a future occasion.

● *What is "stripping"?*

This is a technique used primarily by commercial breeders to remove ("strip") eggs and sperm from selected broodstock. Without the use of this skilled technique, ornamental fish production would be too unpredictable to meet the world demand for particular types of pond fish in predetermined quantities. Stripping of fish should only be carried out by an experienced person.

◀ *This male Sarasa Comet Goldfish* (Carassius auratus *var.*) *is displaying nuptial tubercles on the gill covers and leading edge of the pectoral fin.*

happen in subtropical and tropical areas with species such as Guppies (*Poecilia reticulata*), Mosquito Fish (*Gambusia* spp.) and Medakas (*Oryzias* spp.).

In terms of their overall breeding strategies, fish are classified either as egglayers or as livebearers. As these terms imply, egglayers lay eggs and livebearers give birth to live young. However, such apparently straightforward terms cover a whole spectrum of breeding methods. These range, at one end, from strict egglaying, in which both eggs and sperm are released into the water where fertilization takes place, to internal fertilization and subsequent retention of the eggs within the female's body up to the moment of birth.

All the fish species kept in ponds in temperate zones are typical egglayers. In general, these scatter their eggs among plants and leave both them and the resulting fry to fend for themselves. In subtropical and tropical zones, the range of species that are able to be kept in outdoor ponds throughout the year is far more extensive. So are their breeding strategies, which encompass livebearing (for instance, Guppies), egglaying (as in most other species) and an unusual mixture of both. In this last method (found, for example, among Medakas), eggs can be fertilized internally (but are not always fertilized in this way) and later released to complete their development attached to vegetation.

of spring algal flushes to fish and especially to their breeding cycle (see Selecting and Conditioning Broodstock, page 73). Only in the most severe cases of green water is it essential to clear spring algae.

Breeding Strategies

In some species, such as Koi (*Cyprinus carpio*), Tench (*Tinca tinca*) and Goldfish (*Carassius auratus*), sexual maturity is not attained during the first season, but, once reached, breeding capacity will be maintained for many years. In the case of smaller species with shorter lifespans, breeding can occur after one season or, in certain instances, even within the same season, as can

Sexing Fish

There is no single way of sexing fish; it varies with species. For example, in the case of such species as Guppies, Paradise Fish (*Macropodus opercularis*) and many others, males are more colourful than females. In Guppies, males are also smaller than females, while in Paradise Fish, males are larger and have fuller fins than females.

In the majority of temperate pond fish, notably Goldfish and Koi, the sexes are quite similar to each other, especially outside the breeding season. Even so, males are generally slimmer than females and possess a stronger, often rough-edged, first pectoral (chest) fin spine.

Goldfish and Close Relations

In Goldfish and their close relations, spawning itself can last several hours and usually begins early in the morning. Fish gather either in pairs or, more customarily in a pond environment, in groups that frequently consist of more males than females. Preferred natural spawning sites (all in shallow water) include fine-leaved vegetation like Hornwort (*Ceratophyllum* spp.), fibrous roots, such as those extending into the water from pond-edge plants, and even mats of blanket weed. While many pond keepers favour such spawning media because they are natural, the fact re-

mains that they are not as easy to manage or handle as specially designed spawning mats and brushes. Once loaded with eggs, these have the clear advantage that they can be taken away from the pond environment for hatching in safer surroundings. If, however, the survival of the eggs and young fish is going to be left to chance, then natural vegetation is, obviously, a perfectly good spawning medium.

After much chasing, the pairs or groups of fish end up either side by side or in a bunch in the spawning medium, often with as much as half their bodies above the water surface. Sometimes they can lie virtually motionless for several minutes in this position, especially if the courtship chasing has been especially vigorous. More commonly, there is a good deal of splashing and jostling, during which hundreds or thousands of eggs (depending on species and size) are released and fertilized.

Some of the fish at the end of the queue, as well as juveniles that are not yet ready to spawn, will immediately begin to devour the newly-laid adhesive eggs, which represent a tasty, nutrient-rich meal. Even so, some will invariably escape the onslaught. Such are the laws of nature!

Once spawning is completed – usually after several exhausting hours, involving numerous bouts of egglaying and fertilizing – the fish retire and tend to become quite placid. At this stage, it is always a good idea to take a close look at them, but without disturbing them unduly, to see if any injuries have occurred. Breeding is a hectic time, and fish are liable to get damaged. A few missing scales or fin tears here and there should not cause any great concern, but if major injuries have been inflicted, such as gashes from crashing into planting baskets or the sides of the pond, then it is advisable to remove the injured fish and treat their wounds elsewhere.

● *Is it possible to induce fish to spawn at a predetermined time?*

... Yes, it is. This, in fact, is what happens in many commercial farms, particularly those dealing with food fish, such as carp, or ornamentals like Goldfish or Koi. Induced spawnings, which involve injecting broodfish with carefully calculated amounts of a preparation made from pituitary gland extract, allow breeders to plan their programmes in accordance with market demands and without having to contend with the vagaries of weather or the influences of the seasons. For year-round fish production (which consumer demand makes necessary), brooders are generally kept indoors, where they can be monitored and brought into breeding condition as and when required.

● *How many eggs or fry do fish produce?*

The amount can vary from as few as 20 or so fry in some livebearers, to several hundred eggs in small egglaying species, several thousand in species like Goldfish, and as many as 100,000 per 2.2lb (1kg) of body weight in a fully mature Koi female.

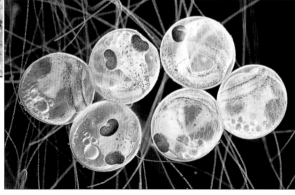

◀ *The Three-spined Stickleback* (Gasterosteus aculeatus) *is widespread throughout Europe and North America. During breeding, the male of this species creates a tunnel on the substratum for its mate to lay eggs in (above left). The male, resplendent in its bright red courting colour, guards the nest and fans water over the eggs until they hatch. (Left) Some eggs of a Three-spined Stickleback; the fish embryos can be clearly seen.*

Other Spawning Strategies

Other egglayers, including the Fathead Minnow (*Pimephales promelas*), do not scatter their eggs but deposit them in a selected, pre-cleaned spot, such as the underside of a lily pad, or a cave, and guard them until they hatch. In Paradise Fish, the site consists of a bubble-nest, which is vigorously defended by the male. In Sticklebacks (*Gasterosteus* sp. and *Pungitius* sp.), the nest is a tunnel built out of vegetation by the male and stuck together with kidney secretions (on the bottom in *Gasterosteus* and higher up among the plants in *Pungitius*).

Livebearers

In livebearing species, the anal (belly) fin of the male is modified into a mating organ (called a gonopodium in the Guppy and fellow members of the family Poeciliidae). During mating, sperm packets (spermatozeugmata) are introduced into the body of the female, where they fertilize the eggs. These then remain in their egg sacs (ovarian follicles) until they hatch a few weeks later and can be released by the female as fully formed fry. Poeciliid females also possess the remarkable ability of being able to store unused spermatozeugmata to fertilize subsequent batches of eggs. A single successful mating between live-bearers can therefore result in several batches of fry being born over the entire breeding season.

Signs of Imminent Breeding

Most temperate pond fish are members of the family Cyprinidae, many of whose male representatives develop in spring what are generally known as nuptial or breeding tubercles. These are small, white pimple-like growths, which, in Goldfish males, are predominantly distributed around the opercular (gill) areas and the front rays of the pectoral (chest) fins. In other cyprinids, such as the Red Shiner (*Cyprinella lutrensis*), they are located on the top of the head and snout, while, in yet others, they appear elsewhere on the body.

In addition to the tubercles, male Goldfish and many other cyprinids do a great deal of chasing (driving) of females. During these chases, males repeatedly nudge females in the abdomen. The

▼ *The breeding strategy of the Golden Medaka* (Oryzias latipes) *is unusual, in that the female carries a "bunch" of between five and fifty fertilized eggs around for some hours, before depositing them on a plant to hatch.*

● *Is it better to stock a pond with more males than females, more females than males, or equal numbers of both?*

A great deal depends on the emphasis being placed on breeding. If breeding is seen as an important factor, then, generally speaking, having more males than females will result in competition between males, thus enhancing one's chances of success. Where breeding is not so important, the sex ratio becomes immaterial. In some fish, such as Koi, females are larger than males and since both sexes are equally colourful, some keepers choose to stock their ponds with more females than males, purely on the basis that larger fish may look more impressive.

● *What is "flock spawning"?*

Flock spawning occurs when fish that do not form strong breeding pair bonds are allowed to mate freely in a pond. This includes species like Orfe, Rudd, Tench, Goldfish and Koi. In flock spawning, groups of fish, rather than pairs, breed simultaneously, thus releasing a large mixture of eggs and sperm with subsequent random fertilization. In most cases, such mixing is unimportant, but where specific traits are desired, individual pairs need to be allowed to spawn away from potential "contaminants". This is, however, difficult to achieve in most ponds.

general impression created during driving is that the females are unwilling participants in this mating game, but closer examination reveals that those females which are really plump with eggs and are ready to spawn seem to allow themselves to be caught quite easily and may even "lead" the males on to a chosen site. It is not merely by chance or sheer persistence that spawning pairs or groups end up among fine-leaved vegetation or strategically placed spawning mats or brushes.

Among egglayers that prepare a spawning site in advance, signs to look for include intensification of colours in the males and hyperactive behaviour concentrated on a restricted area such as a rock, lily pad, or cave, or additionally, as in the case of the Paradise Fish, the appearance of tiny bubbles on the water surface.

In livebearers, constant displaying by males in front of females, followed by mating attempts, are strong signs that young will follow several weeks later. During this "pregnancy", the females will become noticeably fatter.

Hatching and Rearing

Depending on ambient temperature, cyprinid eggs will take between two and seven days to hatch. Among Paradise Fish and other egglayers, hatching can take as little as 24 hours. Over-fast development owing to high temperatures can, however, lead to malformations among the fry.

After hatching, the fry take several days to absorb their yolk sacs and then begin searching for food. Generally speaking, a well-maintained, but not over-filtered, pond will provide a wide array of suitable foods. However, if you are intending to raise a large number of fry, these should be reared away from the pond and fed on either natural pond foods or (far more practicably and safely) on one of the many excellent commercial fry preparations on the market.

Assuming that spawning takes place during spring, young fish will be well enough developed by the end of the season to withstand their first winter as successfully as their adult pondmates.

▲ *Goldfish eggs on a plant, seen 36 hours after laying. To ensure a sizable hatch, many pond keepers use a removable mat or brush for fish spawn.*

Selecting and Conditioning Broodstock

Unless your aim in breeding fish is to improve, or to select, a particular characteristic, then simply choose healthy-looking, attractive fish. Selecting more males than females often results in greater breeding success, as this increases competition.

Fish with obvious deformities, such as out-turned or missing gill covers, a bent spine, etc., should be avoided. Provide the broodstock with a protein-rich diet during the previous autumn, overwinter them properly and maintain good water quality in the pond at all times. Do not be too zealous in eliminating green water in spring (unless water conditions are very poor). In early spring, feed on top-quality foods and offer regular "treats" in the form of deep-frozen, freeze-dried and cultured live foods.

Spawning periods vary with species; for example: Goldfish from early spring through summer; Koi from late spring through summer; Orfe from late spring; Tench from late spring to early summer; Rudd from spring to early summer. It is quite common for spawnings to take place in late summer in temperate regions, particularly if a partial water change has been carried out. However, fish that are born late in the season may not be robust enough to withstand the arrival of harsh winter conditions and should be accommodated in aquaria or indoor ponds until the following spring.

Health Management

THERE IS NO DOUBT THAT PROVIDING FISH with a well-balanced diet will give them considerable resistance to illness. The health of your fish can be harmed either directly by pathogenic organisms, or indirectly by increased susceptibility to environmental stressors that can, in turn, lay them open to disease.

Factors Affecting Fish Health

A good diet, in terms of its composition, does not eliminate all risks. Other factors also play a part. For instance, you may be providing highly nutritious food, but presenting it incorrectly, due to ignorance of the biological characteristics of a particular species. Take the case of the family Cyprinidae, of which many species of ornamental fish, such as Koi (*Cyprinus carpio*) and Goldfish (*Carassius auratus*), are members. Somewhat surprisingly, one of the many characteristics that all cyprinids share is the total lack of a stomach. They do, however, possess a very long gut. A biological consequence of this is that cyprinids are unable to handle large amounts of nutrients in a single feed (the figure is estimated at around 1% of total body weight). Yet these fish need to eat between 2 and 20% of their body weight every day, depending on their age and size. It is thus pointless giving cyprinids one large meal a day and expecting them to grow well and enjoy good health. It is far better to supply them with their daily requirements in a series of small feeds at regular intervals.

Ultimately, it is the degree of disruption of the delicate balance that exists between a fish, its environment and surrounding pathogens that determines whether it will succumb to disease or stay healthy. As well as an adequate diet, good water and regular pond maintenance are also essential. However, even careful attention to all health-related factors cannot guarantee that an outbreak of disease will not occur at one time or other. Nevertheless, preventive steps will reduce the risks considerably.   (*Continued on page 80.*)

Q&A

● *Can fish diseases be transmitted to humans?*

... The vast majority of fish diseases are specific to fish and cannot be transmitted to humans. A few, however – the so-called zoonoses (singular: zoonosis) – can bridge the gap. The most significant zoonotic fish disease is fish TB, caused by at least two *Mycobacterium* species. Although fish TB is not ideally suited for the high-temperature environment that exists within the human body and so has difficulty in spreading, it can, nevertheless, create stubborn, painful, localized cutaneous lesions (skin irritations) which require antibiotic therapy or chemotherapy. Infection with fish TB bacteria is rare, and infection with other bacteria even rarer. Despite this, it is always advisable to take simple, effective precautions by wearing surgical gloves when handling sick fish and by not exposing cuts or abrasions to potential sources of infection, for instance by immersing bare hands and arms in contaminated water.

● *Are there any risks in overdosing fish with medications?*

Medications are designed to be most effective at determined dosage levels. It is for this reason that the capacity of a pond should be calculated as accurately as possible, so that correct dosages can be applied. If fish are over-dosed, there is always the risk that the medication will kill, not just the pathogenic organism, but the fish themselves. Always therefore keep a close watch on treated fish and carry out an immediate partial water change at the earliest signs of distress. Some types of fish, e.g. Orfe (*Leuciscus idus*), are particularly sensitive, thus requiring extra vigilance and, generally, a lower dose of medication.

● *Are there any risks in underdosing fish with medications?*

One risk is that a lower-than-required dose of medication may not clear up the disease, making further treatment necessary. Additionally, low dosage levels that do not kill off the disease-causing organisms can lead to a build-up of resistance and subsequent loss of effectiveness of the medication in question.

COMMON FISH DISEASES

VIRAL INFECTIONS

Carp or Fish Pox

Symptoms: Hard, whitish or cream-coloured, waxy-like patches that may merge to cover substantial areas of the body. Secondary infection can occur if the patches are scraped off.
Treatments: At the moment, viral infections cannot be treated effectively (but some vaccinations have been used and others are under development). Usually, though, viruses are not lethal (but see SVC) and symptoms often disappear spontaneously, particularly as water temperatures warm up. Establishing and maintaining good water quality may help minimize risks.

Lymphocystis

Symptoms: Large cauliflower-like growths (sometimes referred to as tumours) or small, isolated, pearl-shaped nodules or warts, caused by enlargement of affected cells by as much as 100,000 times their normal size, distributed on fins or body.
Treatments: As for Carp or Fish Pox.

Spring Viraemia of Carp (SVC)

Symptoms: Externally: bloated body, blood spots and bleeding under the skin and around the anal area; pale gills. Internally: enlarged spleen and liver; accumulations of fluid; damaged blood vessels.

▲ *A Bubble-eye Goldfish* (Carassius auratus *var.*) *with an advanced case of fin rot.*

▲ *Ulceration in a Common Goldfish following a viral infection.*

Treatments: This is a highly infectious disease which (at least, in some countries), is "notifiable"; that is, the relevant authorities must be informed of an outbreak. There is no known cure and affected fish must be humanely destroyed. Good hygiene and water quality control will help prevent outbreaks. Fish that survive SVC may become immune, but will be potential carriers of the disease.

BACTERIAL INFECTIONS

Fin, Tail or Body Rot (Columnaris Disease)
Flexibacter columnaris and (possibly) other bacteria

Symptoms: Localized or more general attack on body or fins, leading to ulceration or shredding of fin rays. Blood streaks in fin rays usually develop.
Treatments: Acriflavine, Chloramphenicol and other antibiotics*. Furan, Nifurpyrinol and related compounds are also effective, but may be potentially dangerous to humans if their use is prolonged (check with your vet, both for dosage levels of antibiotics and correct use of Furan and related products). Some proprietary remedies are available. Follow instructions to the letter.

Mouth Fungus
Flexibacter columnaris and, possibly,
other bacteria

Symptoms: Whitish growths around the mouth,
gradually progressing into the jaw bones and
anterior cheek area, resulting in actual erosion
of tissues. Note the disease is caused by a bacterial infection not a fungus, despite its common
name.
Treatments: As for Fin Rot.

Bacteraemia (mild) or Septicaemia (acute)
Aeromonas hydrophila

Symptoms: Frayed fins, blood spots, ulcers,

▲ *A severe case of septicaemia, seen here infecting the
body of a Koi.*

▼ *The distending effects of dropsy are clearly evident in
the "pine-cone"-like appearance of this Goldfish.*

Exophthalmia ("Pop-eye"), individually or in
combination.
Treatments: Chloramphenicol, Tetracycline,
Oxolinic Acid (available in medicated flake
form). Some proprietary remedies available.
Chances of cure are variable, depending on
severity of infection. If treatment is being
administered by injection (as antibiotics often
are), ask a vet to carry this out.

Hole-in-the-Head/Body or Ulcer Disease/ Furuncolosis
Aeromonas salmoncida, A. hydrophila as a
secondary agent

Symptoms: Ulcers (often circular) in head
region and/or body.
Treatments: Chloramphenicol, Tetracycline,
Sulphonamides, e.g. Furan-based remedy (see
Fin Rot). Some proprietary remedies available.
After treatment, dress ulcers, if large, with an
oil-based antiseptic cream to protect the wound
and prevent excessive intake of water through
the lesions. Adding salt to the pond water to
give a 3% solution will also help affected fish
to control their water intake.

Fish TB
Mycobacterium spp.

Symptoms: Any one or combination of: loss of
colour/condition, raised scales, Exophthalmia,
loss of appetite, emaciation, ulceration, and
frayed fins.

Treatments: No particularly effective treatment. Some antibiotics, e.g. Terramycin, may work if the infection is not too severe. Fish TB can cause painful localized infected areas in humans. These can prove stubborn to cure. Always observe good personal hygiene procedures, e.g. wear rubber gloves and avoid immersing your hands in water, or handling fish, if you have open/fresh cuts.

Dropsy or Bacterial Haemorrhagic Septicaemia
Aeromonas liquifaciens (ascitae and/or *typica)* and, possibly, other bacteria

Symptoms: Swollen abdomen and raised scales, giving the body a "pine-cone" appearance.
Treatments: Usually no cure, but Chloramphenicol (see vet for instructions) may help. Humane disposal of affected fish, plus good hygiene, are the most effective methods of eradicating the disease and preventing its recurrence.

FUNGAL INFECTIONS

Cottonwool Disease or Fungus
Saprolegnia spp. and *Achyla* spp.

Symptoms: White or cream-coloured fluffy patches on fins and/or body.
Treatments: Proprietary pond remedies widely available, usually based on Phenoxethol, Malachite Green[†], Copper Sulphate, Potassium Permanganate or Methylene Blue. Salt bath: 2–5% solution for 10–15 minutes.

PROTOZOAN INFECTIONS

Whitespot, Ichthyophthiriasis or Ich
Ichthyophthirius multifiliis

Symptoms: Body and fins covered in small white spots; in severe infestations, the spots may appear to join up; fins carried close to the body; violent swimming action or shimmying; scratching against stones, plants and pond equipment.
Treatments: Numerous proprietary remedies are available, many based on Malachite Green[†], Methylene Blue or Copper Sulphate. Repeat treatment is often necessary to attack free-swimming stages of the parasite.

Costiasis or Infectious Turbidity of Gills and Skin
Ichthyobodo (Costia) necatrix

Symptoms: Slimy, bluish-white coloration of the skin and gills; badly affected areas may show blood spots; awkward swimming movements with fins carried close to the body; scratching, as in whitespot.
Treatments: Acriflavine (10ml (0.35 fl oz) of 0.001% stock solution/5 litres (1.1 gallons/1.3 US gallons) of water); salt (about 5g/litre (0.8 oz/gallon)): Formalin (approx. 15ml (0.5 fl oz) of 10% stock solution/5 litres (1.1 gallons/1.3 US gallons) for 15–20 minutes. Since free-swimming stages can only survive for about one hour away from a host, affected ponds can be disinfected by removing fish (not an easy task!) and treating them elsewhere.

Chilodonelliasis
Chilodonella cyprini

Symptoms: Symptoms similar to Costiasis; in addition, the skin in the neck to dorsal fin region may develop a lumpy texture; respiration may be impaired.
Treatments: A combination of Acriflavine (1g/100ml of water) at a temperature of 28°C (82°F) for 10 hours is normally effective; Malachite Green[†] (0.15mg/litre) for an indefinite period. Salt (25g/litre for 10–15 minutes or 10–15g/litre for 20 minutes); Potassium Permanganate (1g/100 litres) for 90 minutes;

▲ *A Three-spined Stickleback* (Gasterosteus aculeatus) *suffering from whitespot.*

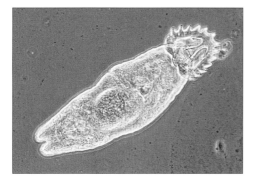

▲ *Skin fluke* (Gyrodactylus) *– one of the most common worm pathogens.*

▶ *A Fish Louse* (Argulus) *attached to the caudal fin of a Goldfish.*

▼ *The swollen abdomen of this Goldfish is likely to have been caused by a worm infestation.*

Methylene Blue (30ml of 1% solution/100 litres (ca. 1.1 fl.oz /22 gallons/26 US gallons) for an indefinite period.

Trichodiniasis
Trichodina spp.

Symptoms: Symptoms similar to Costiasis, but usually less severe. Microscopic examination of the affected fish is necessary for definitive diagnosis.
Treatments: As for Costiasis/Chilodonelliasis.

WORM INFECTIONS

Gill Fluke Infection
Dactylogyrus spp. – Monogenetic Trematodes

Symptoms: Inflamed gills; excessive secretion of mucus on gills; accelerated respiration; gill covers may be held open; scratching of gill covers on pool sides and equipment.
Treatments: Some proprietary remedies are available based on a range of compounds, e.g. Copper Sulphate; salt bath (10–15g/litre for 20 minutes or 25g/litre for 10–15 minutes); Formalin (20ml commercial formalin/100 litres; 0.7 fl.oz/22 imperial gallons/26 US gallons) for 30–35 minutes; Methylene Blue (30ml of 1% stock solution/100 litres) indefinitely.

Skin Fluke Infection
Gyrodactylus spp. – Monogenetic Trematodes

Symptoms: Inflamed patches on the skin, fins and, occasionally, gills; erratic swimming; excessive secretion of body slime; accelerated respiration (if gills are affected); some loss of colour; scratching on pool sides and equipment.
Treatments: As for Gill Fluke.

CRUSTACEAN INFECTIONS

Fish Louse
Argulus spp. – usually *A. foliaceus*

Symptoms: Nervous swimming and jumpiness, frequent, vigorous scratching, often resulting in loss of scales; heavy infestations are accompanied by anaemia and loss of colour; the almost-transparent parasites can be seen attached to the body, mainly along the fin bases (adult parasites can be nearly 0.4in (1cm) long).
Treatments: Organophosphate (Dichlorvos) bath[§] (only recommended for large fish: 1ppm for less than 1 hour, accompanied by very vigorous aeration, as Dichlorvos reduces oxygen tension); Potassium Permanganate (1g/litre for 30–45 seconds or 1g/10 litres (2.2 gallons/2.6 US gallons) for 5–10 minutes); physical removal of individual parasites with forceps followed by disinfection with a proprietary disinfectant, e.g. an Acriflavine-based compound.

Anchor Worm Infection
Lernaea spp.

Symptoms: Long, thin, white, worm-like parasites, up to 0.4in (2cm) in length attached to the body and/or fins; two white egg-sacs at the posterior end are usually visible.
Treatments: Some proprietary remedies are available, e.g. based on Copper Sulphate; Potassium Permanganate – as for Fish Louse. Salt bath (10–15g/litre for 20 minutes or 25g/litre for 10–15 minutes); Trichlorphon[§] has been used (but see below) at 1mg/litre as an indefinite bath, during which it may also dislodge the parasite.

NOTES
Where small doses and volumes of water are quoted, this infers that the treatment should be administered to individual fish isolated in suitable containers, such as large inspection/treatment bowls, hospital tanks or temporary ponds. The same applies to salt bath treatments.

* Whenever antibiotics are being considered, it is wise to check with a vet for accurate dosage levels, even in countries where such products are legally available without prescription.
† Malachite Green is widely available, but this may change, due to its carcinogenic properties (although, at

the levels used to treat fish, there appears to be no evidence of it posing any danger to humans or fish).
§ It is important to check on the legality, or otherwise, of using organophosphates. For instance, the only such product which is licensed for use on fish in the UK (and even then, *not* on ornamentals) is

Dichlorvos. In the UK it is now illegal to sell, or even give away, Trichlorphon products (Trichlorphon breaks down into Dichlorvos, the active ingredient, after a longer or shorter time, depending on ambient conditions), although it is legal to import it for use on one's own animals.

Quarantine

One way of minimizing the risks of introducing diseases into an established healthy pond community is to subject all new fish stocks to a period of quarantine. When stocking a pond for the first time, the pond itself can be used for this purpose, with the fish being kept under observation for a couple of weeks and with further stocks not being bought until you are convinced that the original batch is free of disease or parasites.

Subsequently, all new stocks should be kept under observation well away from the existing population for several weeks, to avoid exposing the resident healthy fish to any diseases that the new fish may be carrying. In practice, it is not always easy to provide separate accommodation for all new fish in the form of either a pond or one or more aquaria. However, every effort should be made to provide such temporary quarters, since just one major "accident" can result in the loss of a whole established fish community, with all the distress that this causes.

Quarantine quarters need not be too elaborate, but should cater for all the basic needs of the new fish (aeration, shelter, and so on). Prefabricated or collapsible paddling pools, large aquaria, or even sheets of butyl rubber or polythene that can be draped over a temporary border made of planks or other suitable materials, can be used. It is important, though, to know the capacity of whatever receptacle is being used so that medication dosage levels can be accurately calculated should treatment become necessary.

Many established pond keepers advocate the use of prophylactic (preventive) medication in the form of a general anti-parasite treatment for all newly acquired fish. Others prefer simply to monitor the new arrivals closely and refrain from adding any treatments (unless absolutely necessary) other than proprietary anti-stress or mucus-protecting conditioners. Still others feed the new fish with pellets or flakes containing antibiotics (usually available under a vet's prescription) during the first couple of weeks.

If it is not possible to provide quarantine quarters, buying quarantined/acclimatized fish from a reputable source and following the guidelines given earlier for selecting healthy fish, their subsequent transportation home and introduction into the pond (see Choosing and Introducing New Fish, pages 56–61), should help avoid all major health risks, although they cannot offer an absolute guarantee.

Whether new fish are transferred from quarantine quarters or from transportation bags into an established pond, extra vigilance is essential for the first few weeks, with action being taken as soon as a problem is observed.

Some Provisos

Fish health is an exhaustive topic that can only be treated broadly here. Detailed information and advice can be obtained from fish health care books, experienced fish keepers and dealers, and – very importantly – your local vet. Some types of fish, such as large Koi, should never be lifted out of the water in a net. For inspection or removal, guide the fish into a floating inspection bowl, using a long-handled pond net. Owing to their large size, such fish can be difficult to handle out of water, for example, for topical application of a medication. For this reason, and to reduce stress levels, larger specimens are often anaesthetized prior to treatment. This procedure is not particularly difficult, but under no circumstances should it be attempted by an inexperienced hobbyist. Either seek the assistance of an experienced colleague, or ask your vet to do it for you. Handling a large pond fish badly (however lovingly and unintentionally one may do it) is, quite simply, not acceptable.

Some general points are also worth stressing. It is, for example, crucial to realize that all medications are stressful, so a careful watch must be kept on the fish being treated and the treatment stopped as soon as any signs of excessive stress, such as loss of balance, are observed. Never be too keen to administer a battery of treatments at the first sign of trouble. This could cause more problems than it solves! All possibilities, such as water conditions, diet, stocking levels, should be checked thoroughly before taking action. Seek advice if there is even the slightest doubt. Both you and your fish will benefit as a result.

The use and availability of some compounds, such as antibiotics and organophosphates, are restricted in certain countries, so it is essential to ascertain the legality or otherwise of treatments with your vet before obtaining supplies. Regulations governing the use of particular medications vary from country to country.

Humane Disposal
of Terminally Sick Fish

Sometimes, despite our best efforts, it becomes evident that a sick fish will not recover. The decision must then be made whether to continue treating it, cease treatment, or bring its misery to a humane end (euthanasia). This is never an easy decision, but it is one that we, as fish carers, are all confronted with sooner or later.

There are several methods of euthanasia, some more acceptable to most pond keepers than others. Chilling and overdosing are thought to be kindest to the fish, and are probably the two easiest methods from the human perspective.

Chilling involves first dissolving two or more tablespoonfuls of salt (depending on the size of the fish) in a bag containing pond water. The fish is then placed in the bag and left in a freezer overnight. It appears that the salt has a calming effect on the fish, whose metabolism gradually slows down as the temperature drops, until it eventually dies.

Overdosing entails adding any of the anaesthetics used for sedating fish for treatment, e.g. quinaldine sulphate, benzocaine or MS222 (at several times the normal dosage level) to a bag or bowl of water containing the sick fish. The fish is then left in the solution until it quietly slips into unconsciousness and eventually dies. While fish anaesthetics may be available without a vet's prescription, it is always advisable to seek expert advice when disposing of a fish.

Q&A...

● *Can the effectiveness of a medication be affected by filters and UV sterilizers?*

Ammonia-adsorbing filter media have little effect on medications. However, other adsorbing media, such as activated charcoal, will remove medication as the treated water passes through the filter. Therefore, if such adsorptive media are present within the pond filter, they should either be removed, or the filter switched off for a few hours. Antibiotics will kill, not just pathogenic bacteria, but also the "good" ones which form part of the filter flora responsible for detoxifying water. Biological filters should therefore be disconnected during whole-pond antibiotic treatment. Ultra-violet sterilizers can neutralize the effects of some medications and should therefore also be switched off during treatment.

● *Are there particularly good or bad times to treat ponds with medications?*

Whether treatment is carried out in hospital tanks/vats, or the pond itself, it is best to avoid both very hot and very cold days, if at all possible. On hot days as temperature rises, so do the metabolic demands of fish for oxygen. Unfortunately, as temperature rises, the concentration of dissolved oxygen available in the water drops. Since, further, a fish's oxygen requirement also tends to rise during treatment, irrespective of the temperature, it is advisable to supply additional aeration while fish are being treated. Cold-weather treatment is inadvisable because, while dissolved

oxygen levels may be adequate, the metabolic rates of both the fish and the pathogenic organisms is slow, thus prolonging the time required for medications to have their full effect.

● *What are anti-stress and mucus-protecting conditioners?*

These are commercial preparations that coat the body of the fish with a protective layer. This layer takes over at least part of the function of the mucus that normally forms a physical barrier between the fish and its environment, but that is often damaged during handling and transportation. By helping to ease a fish into its new pond, anti-stress preparations increase its chances of surviving this critical period.

● *What is the difference between quarantine and acclimatization?*

Strictly speaking, quarantine is a period of isolation lasting 40 days. While a period of this duration may be possible if suitable quarters are available, it is more usual to "quarantine" fish only for as long as it takes to ensure that they are free of disease, i.e. around two weeks. Of course, if any signs of trouble are seen, the period of isolation should be extended. Acclimatization refers to the period that a fish takes to become used to its new environment. It can be as short as the time it takes for new fish to be introduced into a pond following the procedures outlined earlier, or considerably longer – several weeks – depending on how delicate the fish in question is.

Goldfish

FAMILY: CYPRINIDAE
SPECIES: CARASSIUS AURATUS

THE GOLDFISH IS, WITHOUT DOUBT, THE BEST-known fish in the world. It seems that most of the population of the Western world has kept at least one of these delightful fish at some time or other. Countless thousands of children and adults have been introduced to the world of fishkeeping via the once-ubiquitous "fairground goldfish", which can still be won as a prize at some sideshows. Fortunately, this cruel practice of giving away fish in plastic bags has now been banned in many countries and districts. Millions of Goldfish are bred annually in fish farms in many countries. It is a hardy fish, long-lived and undemanding, and extremely cheap as a result of this extensive commercial breeding.

Confronted with the responsibility of having to care for a living creature, many new Goldfish owners go the whole way and buy all the necessary accessories to keep their new acquisitions in the safe, carefully controlled environment they deserve. One thing leads to another and, before they realize it, they have been firmly hooked! A lifetime of dedication, unending interest and fascination then follows.

● *What is the correct scientific name of the Goldfish?*

... Many books still refer to the Goldfish as *Carassius carassius*. This, though, is the scientific name of the Crucian Carp, and has led to much confusion over the years. The Goldfish is also still sometimes referred to as *Carassius auratus auratus*, i.e. as a subspecies of *Carassius auratus*, the other subspecies being the Goldfish's closest relative, *Carassius auratus gibelius*, the Gibel or Prussian Carp. Now the Goldfish and the Gibel/Prussian Carp are each regarded as valid species in their own rights. Therefore, the correct scientific name for the Goldfish is *Carassius auratus*.

● *Why is the Goldfish so called?*

The "gold" in the name refers mainly to the bronze/olive-brown/old-gold colour of the wild-type fish, rather than to the orange of the basic cultivated variety.

Origins of the Goldfish

The history of the Goldfish, which spans more than 1,000 years, is strewn with complications and speculations, which sometimes make it difficult to separate fact from fiction. Poetic references to red-scaled fish in the ancient Western Jin dynasty of China (265–316) are thought to indicate that the first steps towards cultivating Goldfish varieties had already been taken between 1,500 and 1,700 years ago. However, it should be stressed that this early evidence of the fish's existence is by no means conclusive.

The first real piece of documentary evidence comes from the Chinese Sung Dynasty (960–1279), whose records reveal that Goldfish were kept as pets in ponds. Since then, centuries of selective breeding have produced over 100 "official" varieties of Goldfish.

▲ *Among goldfish, wild-type coloration is found not just in true wild specimens, but also during the juvenile stages of many cultivated varieties.*

◄ *These orange-coloured specimens represent the basic form that many aquarists and pondkeepers associate with the "true" Goldfish. They are, however, one step removed from the genuine wild Goldfish.*

Factfile   *Goldfish*

SIZE: Common Goldfish and other slim-bodied varieties can grow up to around 12in (30cm). Fancy varieties are usually considerably smaller, although some exceptionally large Orandas have been bred in aquaria; these are found mainly in the Far East.

DISTRIBUTION: Native to China and parts of Siberia, but has been widely introduced into other areas of the world.

HABITAT: Slow-flowing and still waters in lowland areas, usually where aquatic vegetation is abundant.

TEMPERATURE: From well below 10°C (50°F) to above 30°C (86°F).

DIET: Will accept a wide range of commercially prepared foods, which must include a vegetable component.

BREEDING: An egg-scattering species that displays a distinct appetite for its own eggs. Spawning generally occurs when water temperatures rise above 20°C (68°F) following a cold period. Although the main body of water in temperate-zone ponds may not reach the required level until quite late in spring, shallow areas, for example among pond edge vegetation or above pond shelves, may do so during sunny spring days and this may be sufficient to stimulate spring spawnings. Where this does not happen, spawnings are usually delayed until late spring or early summer.

Popular Goldfish Varieties

Common Goldfish The "classic" short-finned variety, usually orange-coloured, but also available in other single colours and combinations (**1**).

London Shubunkin Single-tailed, short-finned with mottled coloration which should include a degree of blue (**6**).

Bristol Shubunkin Like the London Shubunkin, but with longer fins (**2**).

Comet Slimmer than a Bristol Shubunkin and with a longer tail, preferably pointed.

Jikin Similar to the Common Goldfish, but with an upright double tail.

Fantail Oval-bodied, double-tailed variety with relatively short fins.

Ryukin Superficially similar to the Fantail but with high dorsal (back) fin and very deep body.

Veiltail Similar to the Fantail, but with long flowing fins (**3**).

Oranda Similar to Veiltail, but with raspberry-like growth (called a hood) on the head.

Redcap An Oranda with a white body and red hood.

Moor Double-tailed, but without a hood. The eye may protrude. Body colour: black.

Butterfly Moor Like the Moor, but with shorter tails which are often splayed out and have the appearance of a butterfly.

Pearlscale Oval-bodied, short-finned variety with domed scales that look like pearls.

Hamanishiki Similar to the Pearlscale, but with two "bubbles" on top of the head. Also known as the High-head Pearlscale.

Lionhead Similar to an Oranda, but without a dorsal (back) fin. The dorsal profile is relatively straight.

Ranchu Like the Lionhead, but with a curved back profile (**4**).

Celestial Similar body shape characteristics to the Lionhead, but without a hood and with upward-looking eyes.

Bubble-eye Similar to the Celestial, but with large fluid-filled eye sacs (**5**).

Goldfish Varieties

Selective breeding over many generations has produced countless combinations, some more attractive and long-lasting than others, but all removed, to a greater or lesser extent, from the basic olive-brown coloration of the original *C. auratus*. Indeed, even the Common Goldfish itself is one step removed from the genuine wild type, in that it can be found in a wide variety of colours, but hardly ever the olive-brown of its ancestors, although juvenile specimens often exhibit such pigmentation on a temporary basis. The accompanying table includes most of the best-known varieties of fancy Goldfish. For ease of reference, only a few of the main characteristics have been included. One additional feature, the so-called "Telescope Eye", can occur in some of the double-tailed varieties, giving them an extra characteristic, irrespective of type. There are, for example, Telescope-eyed Moors, Fantails, Ryukins, and so on. Moreover, nostril enlargements that appear almost like cheerleaders' "pom poms" appear on some varieties, most notably Orandas.

In the past, all Moors were, by definition, black. These fish were often referred to as Black Moors, even though, strictly speaking, the word "Black" was superfluous. Nowadays, however, there are Chocolate, Chocolate and Orange and other colour forms which are also regarded as Moors. Although these fish are descended from genuine (black) Moors and therefore

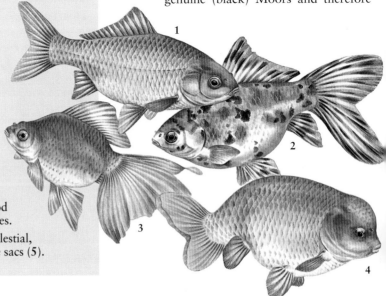

▲ *The traditional Moor gained its name from its overall black coloration. New variants, however, depart significantly from this basic colour.*

▲ *Although some Redcaps have hoods that cover most of the top and sides of their heads, the red coloration in the best specimens is confined to the top of the head.*

carry Moor genes within them, the mere fact that they are not black casts some doubt on the validity of calling them Moors. Perhaps the most sensible approach would be to refer to all black specimens simply as Moors and to qualify the name appropriately when describing one of the other colour types, e.g. Chocolate Moor. One particular variety, a stunning black and white fish that has telescope eyes and a butterfly double tail is, somewhat more reasonably, known as the Panda.

The overall "finish" that a fish exhibit can also vary greatly. Metallic fish contain a considerable amount of the pigment guanine, which gives the body reflective (metallic) qualities, while Matt fish lack the reflective components giving a non-shiny appearance. Nacreous fish have an overall "mother-of-pearl" shine. Calico fish carry a blue

Q & A

● *Is it true that Goldfish don't have any teeth?*

No – although Goldfish do not have lip teeth, i.e. "normal" teeth with which to bite, they do have grinding teeth set farther back in the throat. These are known as pharyngeal teeth.

● *Is there a Goldfish variety called the Meteor?*

The Meteor is a purportedly tailless variety of Goldfish. If it ever existed (and there is considerable doubt about this), it appears to have disappeared completely. Despite having made repeated requests in several countries for anyone who has seen a Meteor, or knows the whereabouts of any specimens, to contact me, no-one has, to date, come forward with any details.

● *Is the Redcap Oranda a true Oranda?*

Yes it is. It is merely a specific type of Oranda, having a white body and a red hood. So, adding the word Oranda to the name is superfluous, since all red-capped fish with a white body are Orandas. A more correct name for this extremely attractive variety is, quite simply, the Redcap.

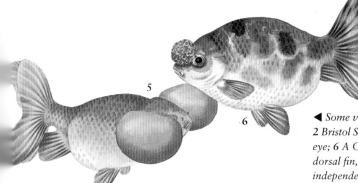

◀ *Some varieties of Goldfish: 1 Common Goldfish; 2 Bristol Shubunkin; 3 Veiltail; 4 Ranchu; 5 Bubble-eye; 6 A Calico "Pom-pon" (this specimen lacks a dorsal fin, but this feature is inherited quite independently).*

ground colour splashed with black, violet, red, brown and yellow. Some specialist Goldfish societies use the Calico classification to include Nacreous and Matt fish as well, as long as they meet the relevant colour criteria.

Pond Goldfish

Only a few types of goldfish can be kept in temperate-zone ponds, owing to the relatively delicate nature of some long-finned and highly developed varieties in relation to low temperatures. The selection is far wider in subtropical regions, but even here, the lack of mobility of some of the fancier varieties needs to be taken into consideration, particularly in ponds that are likely to be visited by predators, whether airborne, such as herons and kingfishers, or swimming, such as terrapins.

Those varieties that possess fluid-filled eye sacs, such as the Bubble-eye and the Celestial, should not be kept either in "predator-prone" ponds or those with rough sides, e.g. cement, which could cause severe injury to these delicate organs. Dark-bodied varieties, such as Moors or

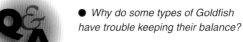

▲ *The striking Sarasa Comet variety is distinguished by a slender body, elongated caudal fin and brilliant red and white coloration.*

Chocolate Orandas, tend to merge with most pond backgrounds and so are probably best reserved for light-coloured pond designs.

Fancy varieties of Goldfish that have round bodies are prone to swimbladder ailments, especially if subjected to sudden temperature fluctuations. If these are avoided, most serious health problems can be prevented.

Q&A...

● *Why do some types of Goldfish have trouble keeping their balance?*

Leaving aside actual diseases of the swimbladder, some perfectly healthy Goldfish exhibit a tendency to float – sometimes upside-down, often after feeding. The fish concerned are usually one or other of the round-bodied varieties (e.g. Fantails, Orandas etc.). Loss of buoyancy control is linked to the shape of the swimbladder which, in such varieties, is shorter and more compressed than in more conventional types. As a result, some round-bodied Goldfish cannot control the passage of gases in and out of the swimbladder with the same precision as their more elongated counterparts can. If, in addition, these fish gulp air at the water surface while feeding, this can result in further, though temporary, loss of balance.

● *In winter, why do round-bodied Goldfish sink and appear to lie awkwardly on the bottom?*

In high-latitude zones with cold winters, all Goldfish (and most other fish) spend much of their time resting on, or near, the bottom of the pond. While normally-shaped fish can keep upright without any trouble at such times, round-bodied varieties can experience some difficulty as the volume of gas within their swimbladder contracts with falling water temperatures. When this happens, they end up lying on their sides or at an angle to the vertical. Once temperatures begin to rise, they regain their control and right themselves, apparently with no ill effects. Nevertheless, it is wise to transfer round-bodied Goldfish indoors during the winter in areas where such conditions are likely to occur.

● *Is it safe to leave Goldfish in a pond over winter when it freezes?*

Yes it is, as long as the pond depth is such that there is a sizable unfrozen layer below the ice and as long as the variety is hardy enough (many fancy types are not).

● *Can I put Goldfish in my wildlife pond?*

Strictly speaking, only wild-type Goldfish are suitable for the wildlife pond. However, orange-coloured Common Goldfish are also usually deemed acceptable.

● *How can swimbladder problems be avoided or minimized?*

Although floating or sinking may appear distressing, it does not normally seem to bother the affected fish unduly. Two useful ways of minimizing the incidence and severity of swimbladder malfunction are to avoid abrupt water temperature changes and to release presoaked food under the surface of the water at feeding time.

Koi

FAMILY: CYPRINIDAE
SPECIES: CYPRINUS CARPIO

EVERY POND OWNER SEEMS TO WANT TO OWN Koi. At least, this is broadly true of today's pond keepers. However, this hasn't always been the case. Even as recently as the early to mid-1980s, Koi keeping tended to be regarded as a rather exclusive sector of the pond hobby, because the fish were so expensive. Many potential owners were also put off by the names used to refer to the different varieties of these glorious fish. It seemed to some as if you needed to learn Japanese before you could attempt to keep Koi!

Changing Scene
The world of Koi keeping has changed dramatically in recent years. Certainly, the high-quality, very expensive fish still exist – and deservedly so. Also, Japan still leads the world when it comes to the top Koi bloodlines. But other countries, such as Israel, the United Kingdom and the United States, are now producing huge numbers of Koi as well.

Some of these fish may not be as "pure" in terms of their lineage as those from a long-established Japanese bloodline, but they are highly colourful and every bit as robust as pedigree stock. One distinct advantage is that they are considerably less expensive than pedigree fish, which command very high prices. This expansion of the market has also resulted in the appearance of a wider variety of competitively priced Japanese Koi than ever before.

The upshot of these changes is that Koi keeping has become much more accessible for the general pond keeper. This, in turn, has stimulated a massive growth of interest in these fish from pond owners who would not have considered keeping them in the past. Modern Koi keepers cover a broad spectrum, from those who are

▼ *Nowadays, Koi can be obtained in a wide variety of colours, patterns, sizes and quality to satisfy all tastes and to suit all budgets.*

Factfile *Koi*

SIZE: Wild carp reach 39–40in (1m). Most Koi are somewhat smaller. Koi measuring above 28in (72cm) are officially regarded as Jumbo Koi.

DISTRIBUTION: *Cyprinus carpio* originated in the Danube and other rivers of the Black Sea basin, but is now found throughout Europe (except the far north) and elsewhere. Barring some escapes or introductions, Koi are not found in the wild.

HABITAT: Carp prefer sluggish or still waters, especially warm ponds with abundant vegetation. Koi tolerate a wide variety of conditions, from muddy natural ponds to clinically clean Koi pools.

TEMPERATURE: From well below 10°C (50°F) to above 30°C (86°F) – but not for extended periods at the higher temperature.

DIET: A wide range of foods, which must regularly include vegetables. Koi especially like regular treats (e.g. prawns, brown bread, lettuce, garden peas. They cannot digest the outer skin of peas, so crush them and spit the skin out.

BREEDING: An egg-scattering species which will eat its own eggs. Spawning usually begins at a water temperature of around 20°C (68°F), but can occur at temperatures as low as 17°C (63°F). Spawning activity is very vigorous and can result in injuries to the fish. The use of spawning brushes (ropes) or mats, or the construction of a large floating net "cage" in which selected brood fish can be isolated, is strongly recommended. Eggs hatch in 3–4 days at temperatures between 20–22°C (68–72°F).

perfectly happy to own just non-pedigree fish, as long as they are attractive and healthy, to confirmed specialists who only keep pedigree Koi, or even pedigree specimens of just one variety of Koi, such as the *Kohaku*. The majority of Koi keepers, though, have a preference for a mixture of fish, which may vary in "quality" with regard to their genetic make-up, but which are all equally healthy and vigorous.

The First Koi

The first fish that were genuinely regarded as ornamental varieties of carp were developed in the Niigata Prefecture on the Japanese island of Honshu during the early part of the 19th century. The three main colours usually reported for these early Koi are red, white and yellow. It is probable that these coloured fish, which were subsequently bred into other forms, originally

arose purely accidentally as the result of sponta-
neous mutations affecting the colour genes in
wild-type Common Carp stocks. There are accounts of coloured carp (red and grey) from an
earlier period (around 530 BC), but, since these
Chinese fish were, almost certainly, only kept as
food fish, it would be misleading to consider
them as the true original Koi in our generally
accepted interpretation of the term.

Biological Identity

In the relatively short time that Koi have been
kept, this majestic fish has been bred into so
many configurations that it is easy to lose sight
of what, biologically speaking, a Koi actually is.
Looking at today's spectacular Koi varieties, it is

difficult to believe that they are all descended
from what some people regard as a dull, drab
fish. To others, however, the Common Carp is,
in its own way, every bit as special a fish as
its jewelled descendants. Apart from Koi, the
other types of *C. carpio* recognized are: Common Carp, the "basic" wild-type, which is fully
scaled; Leather Carp, a scale-less or naked variety; and Mirror Carp, an almost scale-less variety with a few rows of large, reflective scales. All
of these characteristics have been bred into Koi
varieties over the years, giving rise to some
amazing combinations.

Other characteristics of *C. carpio* that can still
be seen in its cultivated descendants, as well as
Koi, include a maximum body length of around

◀ *A small example of an* Asagi *Koi. These elegant and
subtly coloured fish are greatly favoured by devotees of
blue/red-based Koi.*

▼ *Young Koi such as these can be kept in well-filtered
and efficiently managed aquaria, but need to be
transferred to a pond as soon as they begin to grow
towards their full adult size.*

Major Koi Categories

Non-Metallic Varieties

Kohaku White base colour with red markings.

Tancho Sanke (or *Sanke*) White base colour with superimposed red and black markings.

Showa Sanshoku (usually referred to as *Showa*) Black base colour with red and white markings.

Bekko Black markings on a white, red or yellow base colour.

Utsurimono White, red or yellow markings on a black base.

Asagi Blue base with red on the belly, fins and sides of head; top of head is light greyish-blue.

Shusui Similar to the *Asagi*, but with dark blue coloration restricted to the large *Doitsu* scales; naked areas are light blue, except for the head, which is light greyish-blue.

Koromo White base colour with red markings that are over-laid with darker patterning.

Kawarimono All other types of non-metallic Koi, including single-coloured fish, e.g. *Ki-goi* (yellow fish), almost single-coloured fish, e.g. *Hajiro* (black fish with white fin edges) and *Goshiki* (five-coloured fish – but see right).

Tancho Single red marking on the top of the head; this classification is reserved for *Kohaku*, *Sanke* and *Showa* Koi that bear this very distinctive head spot.

Metallic Varieties

Hikarimono (or *Ogon*) Single-coloured fish.

Hikari-Utsurimono Metallic versions of *Utsuri* and *Showa*.

Hikarimoyo-mono Term embracing all other types of metallic Koi.

Other Categories

Kinginrin Fish having 20 or more reflective/sparkling scales; previously, judging of this category at Koi shows embraced fish of any variety, provided they had the appropriate scalation. Nowadays, only *Kohaku*, *Sanke* and *Showa* are included.

Goshiki Five-coloured Koi classified within the *Kawarimono* in countries other than Japan.

Doitsu *Doitsu*-scaled fish (see main text) are generally classified under their colour variety. However, in Japan, *Doitsu* fish of any variety are classified separately, except for *Shusui*.

● *Do Koi and Common Carp have any close relatives?*

... Owing, possibly, to the extremely wide distribution of the Common Carp, the genus *Cyprinus* is often regarded as being monotypic, i.e. as containing the single species *Cyprinus carpio*. In fact, it also contains other, lesser known (and less widely distributed) representatives, such as *C. exophthalmus* and *C. yunnanensis*.

● *What are Fairy or Butterfly Koi?*

Fairy or Butterfly Koi are long-finned fish of any variety. They are especially popular among newcomers to Koi keeping. Fairy Koi have been around for many years, but it is only since the early 1990s that they have been available in any significant numbers.

● *Do Koi lose their teeth?*

Yes. These teeth are not like ordinary teeth, though. Koi, just as their close cousin, the Goldfish, do not possess "lip" teeth. The ones we occasionally find at the bottom of our ponds are pharyngeal or "throat" teeth, which are used for grinding food and which can become dislodged (and replaced) from time to time.

40in (1m), four barbels (two on each side – on the upper lip and the corners of the mouth) and a dorsal fin with a long base that stretches over a considerable length of the back of the fish, containing 17–22 branched rays, preceded by a strong, serrated (toothed) spine at the front.

Koi or Koi Carp?

In virtually every country where Nishikigoi (Koi for short) are kept, many people call them "Koi Carp". The use of this latter label has become so widespread that it is in danger of becoming accepted, particularly by newcomers to Koi keeping, as being correct – but it is not!

No-one would dream of referring to a Goldfish as a "Goldfish Fish". Yet, to call Koi "Koi Carp" amounts to the same thing. "Nishikigoi" means "Brocaded Carp", i.e. the word "Carp" is already encompassed by the term (and, when shortened to "Koi", we are also subsuming the word "Carp"). Consequently, if we refer to a Koi as a "Koi Carp", this is tantamount to calling it a "Brocaded Carp Carp"!

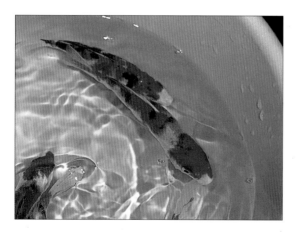

▲ *The traditional bright blue of the Koi viewing bowl is designed to show off the remarkable colours of these fish to best effect.*

Koi Classification

The finer points of Koi classification are highly specialized and are of major significance, particularly to anyone wishing to establish a quality (pedigree) collection, or enter fish in competition. The vast majority of such enthusiasts will therefore construct their water schemes with the Koi as the undisputed central focus of attention, often to the total exclusion of plants or, sometimes, other types of fish.

This degree of specialization requires detailed knowledge, not just of what constitutes, say, a *Kohaku* (a red and white Koi), but of the different types of *Kohaku,* and even the specific bloodlines of such fish. Anyone requiring this type of information is strongly advised to seek out specialist Koi literature and join one of the many aquatic societies dedicated exclusively to Koi keeping, breeding and showing. For the more general pond keeper or water gardener, the details given here and in the accompanying table should prove sufficient to distinguish the main types of Koi, thus allowing informed choices to be made.

In broad terms, Koi are classified according to their colour, pattern of markings and scalation. Within these three criteria, though, there are numerous permutations, resulting in well over 100 recognized varieties. Colours, for instance, are referred to by the appropriate Japanese word, such as *hi* (meaning red), *sumi* (meaning black), *ki* (meaning yellow), and so on. Therefore, when referring to a fish in which the red markings are particularly good, one would say that its *hi* is of good quality. Patterns are also referred to in Japanese. For example, a "lightning" or zig-zag pattern is referred to as *Inazuma* patterning, while a *Tancho* pattern tells us that the fish in question has a red patch in the centre of the head, and so on.

Scalation can be either complete (metallic), or restricted (or almost absent) as in Leather and Mirror Carp, these latter being referred to as having *Doitsu* scaling. Other forms of scalation include highly reflective/sparkling types (*kinginrin* or *ginrin*), or *fucarin*, in which the scales are small, and show the skin in between.

Along with overall body colouring, these various features have been classified over the years into a series of major categories, with innumerable subcategories. While opinions may differ on just how many such groups there are, the accompanying table incorporates all the main ones, including two (*Goshiki* and *Doitsu*) that are recognized as separate categories, at least in Japan.

Longevity

There has been considerable controversy about this subject for many years. It was long accepted that Koi could live for over 200 years. This belief was based on the case of a renowned female Koi named Hanako, which died in 1977 and was thought to be 226 years old. Her age was estimated by counting the number of growth rings in her scales. Koi, like all other fish that live for more than one year, lay down rings of clearly discernible and differently coloured material in their scales and otoliths ("ear" bones) as they grow. As growth slows down during the colder months of the year, the band of hard scale tissue secreted is dense, while at other times it is lighter and wider. Theoretically, then, finding the age is simply a matter of counting the rings. However, factors other than the seasons (e.g. disease and stress) can occasion the growth of scale tissue, which makes calculating the age far more difficult. By this reckoning, an age of 226 is almost certainly an overestimate. An age of around 50 years for a Koi would, however, be regarded as very respectable by any standard.

● *Is it true that Koi don't have a stomach?*

... Koi, like other members of the family Cyprinidae, whose natural diet consists mainly of vegetables, do not possess a distinct stomach like flesh-eating species do. Instead, the structure of their alimentary canal is more generalized, and assimilation of nutrients therefore occurs along an extended length of the gut. (See also Foods and Feeding, page 65).

● *How many eggs can a mature Koi female produce?*

Although several factors, for example diet or state of health, can affect the total number of eggs that a female Koi can produce, a good approximate figure is 45,000 per lb (about 100,000 per kg) of the female's body weight. Spawnings of well over a quarter of a million eggs are thus quite commonplace.

● *What is the best food to offer Koi?*

Koi will accept a wide range of foods. Some specialist keepers prefer to mix their own formulations, which consist of animal- and vegetable-based components. However, modern prepared Koi diets are so nutritious, well-balanced and varied that there really is no need to resort to home-made recipes. There is a wide range of commercial preparations available for Koi of differing ages and sizes, for conditioning broodstock and even for the changing seasons, so the best advice is to visit an aquatic centre and spend some time over your choice. If in doubt, ask for assistance or contact the various manufacturers direct – their details appear on the packaging.

● *I have seen Ghost Koi for sale. What are these, and what conditions do they need?*

There are a number of different opinions on what constitutes a Ghost Koi. One of the more widely held views is that Ghost Koi were first developed from a crossing between a *Sanke* male and a Mirror Carp female. The resulting offspring did not possess the distinctive colours of the *Sanke*, but had either Common Carp characteristics, or shiny scales and a pronounced dark marking on the top of the head. Whatever the precise details of their ancestry, Ghost Koi can vary in colour from creamy white (often with very attractive dark scale edgings), to a very dark coloration similar to that of the Common Carp. They grow vigorously and are exceptionally hardy, and so require no special conditions.

Stocking Your Pond

Koi can be bought at any time of year. However, the best pedigree stock from Japan is usually available from November onwards for several months, following the October harvest of Koi.

If your pond is new, then buy only a fraction of your eventual stock, and build this up gradually over a period of a few months. This allows the filtration system to keep up with the gradually increasing load of fish wastes.

As a general rule, follow the same guidelines as for other pond fish, allowing around 24sq in (150sq cm) of surface per 1in (2.5cm) length of fish, excluding tail. Remember that a reasonable density of young, smallish fish can very soon become an overpopulation of larger ones, with all the problems that this can create. Therefore, stock cautiously at all times and install the most efficient water treatment system you can afford. This will give you a greater safety margin.

▼ *Head of a Ghost Koi, showing clearly the characteristic barbels that help identify it as a descendant of the Common Carp* (Cyprinus carpio).

Other Cyprinids FAMILY: CYPRINIDAE

THE MOST WELL-KNOWN CYPRINIDS ARE THE Goldfish (*Carassius auratus*) and Koi (*Cyprinus carpio*). However, a number of other related species are now widely available for garden pond enthusiasts, which add an extra dimension of variety and excitement.

Orfe

Leuciscus idus. Also known as the Ide, this is a slender, silver-bodied fish which, in its wild form, is best suited to wildlife ponds. The gold-coloured variety – fittingly known as the Golden Orfe – is a better choice for ornamental ponds. In recent years, other colour types have been bred, such as the Blue Orfe and the Marbled Orfe, but neither has overtaken the Golden Orfe in popularity. Both are still attractive fish in their own right, but the dark coloration of the Blue Orfe does not always look its best against a pond background, especially when seen from above, while the light patches of the Marbled Orfe are not as intensely coloured as in the Golden Orfe.

Q&A...

● *Can Orfe be kept with Goldfish and Koi?*

Orfe are tolerant of other species and can therefore be kept with a wide selection of pondmates. However, being predatory by nature, they will consume livefood, which can include very young fish.

● *Are the cultivated varieties of Orfe hardy enough to withstand winter conditions in temperate zones?*

Leuciscus idus is a hardy species, despite its sensitivity to certain medications and its high oxygen requirements. It will therefore be able to withstand temperate winters, providing an adequate diet has been supplied during autumn, the water quality is good (and oxygen-rich) and the pond is deep enough to accommodate large specimens comfortably (18in/46cm depth should be regarded as an absolute minimum, but even this is too shallow in regions that experience very cold winters).

▼ *Many different types of cyprinid can co-exist quite happily in the same pond, especially if they are of similar sizes. Shown here are two gold-coloured Koi (Cyprinus carpio – top centre, in the foreground, and bottom right), some Golden Orfe (Leuciscus idus – e.g. below and slightly behind the top Koi), a Golden Rudd (Scardinius erythrophthalmus – between and slightly in front of the Koi and Orfe) and a Golden Tench (Tinca tinca – below the central Orfe, but pointing in the opposite direction).*

Orfe are active shoaling fish that prefer the surface areas of ponds and are therefore always visible – an extremely obliging aspect of behaviour from the human point of view. However, this surface shoaling is only strongly evident when a number of specimens are kept together. Single specimens will frequently associate with other types of fish, but unless these, too, are surface or midwater swimmers, like Rudd (*Scardinius erythrophthalmus*), single fish will often become timid and spend much of their time among the shelter of the pond vegetation.

Orfe are usually sold as small specimens (3–4in/7.5–10cm). Eventually, they will grow into substantial fish measuring 18in (46cm) or more. They therefore need a large pond if they are to develop their full potential. Orfe require oxygen-rich water and can suffer when supplies run low (as during overcast, muggy weather). Supplementary aeration should be provided at such times by means of airstones, venturis (aerating devices), fountains/cascades, or even

Factfile *Orfe*

SIZE: Grow up to about 39–40in (1m) in length in wild. Cultivated varieties do not often exceed 24in (61cm).

DISTRIBUTION: Native to the River Danube basin and most parts of northern Europe (but not Norway). Has been introduced into many European countries, but appears to be absent from southern regions. There is at least one established population in Maine, USA.

HABITAT: Lives in large shoals in lowland, flowing waters and lakes. Also known from some brackish water areas.

TEMPERATURE: From near-freezing to over 30°C (86°F).

DIET: While being primarily a predatory species, it will, nevertheless, accept a wide range of commercial pond foods.

BREEDING: Eggs are scattered among fine-leaved vegetation or on spawning mats/brushes in late spring and can take as long as 20 days to hatch.

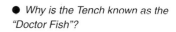

● *Why is the Tench known as the "Doctor Fish"?*

A... It was once believed that the rich mucus layer that covers the body of the Tench, and which makes it feel very slimy to the touch, possessed medicinal qualities. For example, it was claimed that if injured fish rubbed their wounds against the body mucus of a Tench, this would aid healing. In the Middle Ages, Tench body slime was also used to treat human ailments as diverse as jaundice and toothache. There is no scientific evidence to support any of these superstitions, and today the term "Doctor Fish" is hardly, if ever, used in connection with Tench.

● *Do Tench pose any danger to other pond fish?*

Not directly, since Tench are peaceful community fish. Indirectly, though, they can constitute a health risk, but only if the pond in question is poorly maintained. Because Tench are bottom-foraging fish, they tend to churn up considerable amounts of debris. If this debris is excessive – and if it consists of poorly-oxygenated sediments – it can compromise the water quality, which, in turn, can create a health risk for more delicate pondmates. However, in well-maintained set-ups with little bottom debris, Tench will pose no threat whatsoever.

● *How many Tench should I buy?*

Your stocking level will be determined by pond size and personal preference, but, generally speaking, most pond owners buy only 2–4 specimens.

▲ *The full beauty of the blue-coloured morph of the Orfe or Ide* (Leuciscus idus) *is best appreciated when the fish is seen in side view.*

by directing a fine water spray from a garden hose onto the surface of the pond. Orfe are also sensitive to some medications, so check directions and dosage levels carefully before administering any treatment.

Tench

Tinca tinca. The Tench was obtainable in two basic colour forms for many years: the genuine wild type (Green Tench) and a cultivated variety known as the Golden Tench. In the early 1990s, a sprinkling of other colour forms started to appear in aquatic stores, most notably a variety with deep-orange/almost-red coloration, which was sold as the Red Tench, and a deep orange-red and white form, which was labelled the Red-and-White Tench. Surprisingly, it would appear that neither of these two striking varieties has yet become available in sufficient quantities to make a significant impact on the hobby.

Of the two traditional varieties, the Green Tench is more genuinely suited to wildlife ponds than its golden counterpart. Both are bottom-dwelling fish and this makes the former hard to see, except at feeding time. For ornamental ponds, the Golden Tench (with or without dark speckling) is therefore a better choice, since it is more visible, even when resting on, or swimming along, the bottom.

Tench are shoaling fish that are often sold as small specimens of 4–6in (10–18cm) length. Growth is relatively rapid under favourable conditions; bearing in mind the large size attained by this species, only spacious ponds should be regarded as suitable accommodation for mature specimens. Male Tench can be easily identified, even outside the breeding season, by their pelvic (hip) fins, which are much larger than those of females. Males are also generally far less heavily built fish.

In marked contrast to Orfe, Tench are undemanding in their requirements, being able to tolerate low oxygen levels, highly acid (peaty) waters and a wide range of other unfavourable environmental conditions. The Green Tench is also able to change its colour (to a certain

▲ *A Green Tench* (Tinca tinca). *It is acceptable to house small specimens of Tench in aquaria, but they must eventually be transferred to more spacious accommodation.*

Factfile   *Tench*

SIZE: Wild specimens measuring as much as 28in (71cm) have been reported. Usually, in ponds, a size of 12–16in (30–41cm) is attainable in about three years under favourable conditions.

DISTRIBUTION: Widespread in Europe and stretching east to Russia. It is, however, absent in northern Scandinavia, possibly Scotland, Ireland, Iberia and the Adriatic coastline. It has been widely introduced into regions outside its original natural range, including a number of locations in some of the above areas and the US, where it was first released in 1880.

HABITAT: Tench like muddy bottoms in which they occasionally bury themselves during particularly cold spells. Slow-flowing or still waters are preferred, particularly those with abundant aquatic vegetation.

TEMPERATURE: From near-freezing to over 30°C (86°F).

DIET: An omnivorous species that will therefore eat a wide range of both animal- and plant-based foods.

BREEDING: Adhesive eggs (between 300,000 and 900,000) are laid among plants by mature females in late spring and early summer. Hatching takes about 6–8 days, depending on pond temperature.

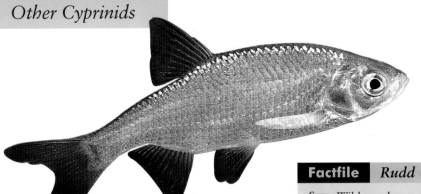

degree) in tune with its surroundings, becoming lighter with golden-orange lips in shallow water that contains little vegetation, and considerably darker in deep, heavily vegetated waters.

Rudd

Scardinius erythrophthalmus. The Rudd or Red-eye inhabits the midwater and surface regions of the pond. It occurs in more than one colour form: the wild type, which has silver coloration; and the Golden Rudd, which is metallic orange/yellow on the top of its body, with a pronounced reddish tinge to its fins. The wild type is a better choice for wildlife ponds, while the Golden Rudd is more suited to ornamental ponds. Both forms are omnivorous and are particularly partial to surface insects.

Rudd are peaceful shoalers which do well in the company of other pond fish, often associating with Orfe if either species is kept singly or in pairs, rather than as a group. There is a population of Rudd in Romania that has adapted to the very specific conditions that exist in some hot springs found in the west of the country. The temperature of the water in this locality ranges from 28–34°C (82–93°F). These fish, which are said to die if the water temperature drops below 20°C (68°F), are usually regarded as constituting a subspecies: *S. e. racovitzai*. As far as is known, no specimens have ever been introduced for ornamental purposes into any country.

The wild form of the Rudd is sometimes confused with the Roach (*Rutilus rutilus*), which has a similar appearance. However, there are certain features, such as eye colour, body shape and fin position, that distinguish the two species.

Factfile — Rudd

SIZE: Wild-caught specimens up to 18in (46cm) reported. Cultivated types are usually smaller.

DISTRIBUTION: Widely distributed in Europe north of the Pyrenees, but absent from northern Scotland and much of Scandinavia. Introduced into a number of exotic locations, including several in the US, such as the lower Hudson River drainage, New York and Maine.

HABITAT: Slow-flowing and still lowland waters, preferably heavily overgrown with waterside and submerged vegetation.

TEMPERATURE: From close to freezing point up to a maximum of 34°C (93°F; for the Romanian population).

DIET: Omnivorous; Rudd will eat a wide range of animal- and plant-based pond foods.

BREEDING: As many as 200,000 adhesive eggs can be laid by large females from spring to early summer. Hatching takes between 8–15 days, depending on temperature.

Grass Carp

Ctenopharyngodon idella. Grass Carp or White Amur were almost unknown among pond keepers until the 1980s, when their reputation as avid consumers of vegetation encouraged their use in ponds as a natural/biological means of weed control. Few, if any, species of submerged pond plants can survive the close attentions of a shoal of Grass Carp. In the main, they are kept by Koi enthusiasts, whose large plantless pools can accommodate these substantial fish.

Most Grass Carp kept in ornamental ponds are albinos. These appear particularly attractive against the dark background and crystal-clear water that are characteristic of Koi pools. Wild-type Grass Carp represent a better choice for wildlife ponds, but, since they will eat succulent underwater vegetation, this makes them unsuitable for certain types of natural scheme.

● *What is the difference between Roach and Rudd?*

... Although they are superficially similar, Rudd and Roach (*Rutilus rutilus*) can be told apart from one another relatively easily. The most distinctive feature is the positioning of the dorsal fin in relation to the pelvics. In Rudd, the dorsal fin is set farther back along the body than the pelvics. In Roach, they are above each other. Other distinctive features relate to details of coloration and body shape. Rudd are less silvery (more bronze-coloured) than Roach and, despite the fact that one of its common names is "Red-eye", the Rudd's eyes are in fact orange-red, while the Roach has genuinely red eyes. Rudd have deeper bodies and their finnage (particularly the pelvic and anal fins) are redder than in Roach. Finally, the mouth slopes upwards in Rudd, indicating that this species feeds mainly from the surface or in midwater. In Roach, the mouth position is more terminal.

● *Can Rudd and Roach interbreed?*

Yes they can. Such hybrids have redder fins than Roach and orange-yellow eyes like Rudd. Rudd also hybridize, though less frequently, with Bream (*Abramis brama*), White or Silver Bream (*Blicca bjoerkna*) and Bleak or Alburn (*Alburnus alburnus*). Hybrids are notoriously difficult to identify and most are infertile.

● *Can Rudd be kept safely with small fish?*

As a general rule, Rudd are peaceful towards all other pond fish. Having said this, they do have some predatory instincts, but these are only evident when large individuals are kept in the company of very small specimens of other species.

● *Are all silver-coloured Rudd wild-caught?*

No. Even where it is permitted to collect this species from the wild, the large majority of Silver Rudd that are offered for sale are farm-bred specimens. However, this type of Rudd is much less frequently encountered than the golden variety.

Factfile *Grass Carp*

SIZE: Up to 50in (1.25m) reported in wild specimens. Growth is extremely rapid. The albino variety may be slightly smaller at full size, but is nevertheless still a substantial fish.

DISTRIBUTION: The Grass Carp originated in the River Amur and the surrounding floodplain in China. It was later introduced into parts of eastern Europe and, gradually, into more western countries. It is also now present in over 30 states in the US.

HABITAT: Flowing bodies of water with abundant vegetation are preferred, including the backwaters of larger rivers. Despite this, Grass Carp do well even in well-filtered Koi pools.

TEMPERATURE: From below 10°C (50°F) to c.25°C (77°F); higher for breeding.

DIET: Herbivorous; will consume large amounts of a wide variety of plants. Where these are not available in sufficient quantity (as in most ponds), the diet offered must include a vegetable component. They will even accept lawn shavings, as long as they do not contain toxic substances.

BREEDING: Many thousands of eggs are released into the water, from where – in the wild – they are carried downstream and hatch into fry that will leave the main channel and move into backwaters and floodplain areas. As the temperature at which Grass Carp spawn is around 27–29°C (81–84°F), they are unlikely to breed in most temperate-zone ponds.

▼ *The Grass Carp (Ctenopharyngodon idella) has a wide mouth especially adapted for plant feeding. Its coloration makes the albino form far more popular among pond keepers than the wild type.*

Minnows, Shiners and Related Species

The term "minnow" is generally thought to include many fish that are not always referred to as "minnows", but are closely related to them. Most notable among these are various "dace" and "shiners" of American origin. There are numerous species that might be considered suitable for ponds, particularly in tropical, subtropical and (with appropriate caution) temperate zones that do not experience severe winters. Those treated here are the best-known examples in subtropical and temperate areas.

Minnows were not traditionally regarded as pond fish. One of the main reasons for this is that their colours are usually seen to better effect from the side than from above; for example, the Southern Red-bellied Dace (*Phoxinus erythrogaster*) develops a red ventral region. This is totally lost in a pond, but can be fully appreciated in an aquarium. Another factor was their relatively small size, with some species like the White Cloud Mountain Minnow (*Tanichthys albonubes*) growing no larger than about 1.8in (4.5cm) and some of the bigger ones not exceeding 4in (10cm).

Today, however, minnows, shiners and dace (other than the "true" Dace, *Leuciscus leuciscus*) have become more popular among pond keepers.

A number of factors have contributed to this change in attitude. Firstly, in the 1980s, the golden morph of the Fathead Minnow (*Pimephales promelas*) from the US began to be offered for sale in Europe as an ornamental aquarium fish. Originally it was cultured commercially in the US as live bait in sport angling. This beautiful, hardy, active, small shoaling and easy-to-breed species stimulated an interest in similar species that had largely lain dormant, or had simply not existed, until then. The result was an upsurge in demand for other similar coldwater fish, especially those that could also be kept in tropical aquaria. Of these, the two most popular species were the Red Shiner (*Cyprinella lutrensis*) and the Southern Red-bellied Dace, the latter rapidly outstripping the European Minnow (*P. phoxinus*), but not the old favourite, the White Cloud Mountain Minnow, in popularity among coldwater aquarists. As water gardening expanded rapidly during the 1980s, so the interest in keeping these fish species spread to pond keepers. This trend was promoted by garden centres (at least, those in European countries) expanding their aquatic sections to include aquarium fish.

Finally, with the growth in environmental awareness among animal and plant hobbyists, colour was no longer invariably considered the

---

Some Minnow Species Suitable for Ponds

	maximum adult size

Fathead Minnow　　　　　　4in (10cm)
(Rosy Red, Golden Minnow)
Pimephales promelas
Distribution: Most of North America, including many exotic locations as far south as Mexico
Temperature: Approximately 10–25°C (50–77°F)
Eggs stuck onto roof of cave or other equivalent

Southern Red-bellied Dace　　3.5in (9cm)
Phoxinus erythrogaster
Distribution: From New York State, westwards to Minnesota, including the Great Lakes and the Mississippi River basins
Temperature: Approximately 10–25°C (50–77°F)
Eggs scattered over substratum

European Minnow　　　　4.7in (12cm)
Phoxinus phoxinus
Distribution: Most of Europe, but absent from Norway and most of Greece, Italy and Spain

Temperature: Below 10°C (50°F) – c.22°C (72°F)
Eggs scattered over substratum

White Cloud Mountain Minnow　1.8in (4.5cm)
Tanichthys albonubes
Distribution: One type is found around Canton (China) and the other around Hong Kong
Temperature: Below 15°C (59°F) – c.25°C (77°F)
Eggs scattered among vegetation

Red Shiner　　　　　　3.5in (9cm)
(Sandpaper Shiner, Red Horse Minnow, Rainbow Dace)
Cyprinella lutrensis
Distribution: Mississippi River basin and Gulf drainages west of the Mississippi - widely introduced into other locations, including northern Mexico
Temperature: Approximately 10–25°C (50–77°F)
Eggs deposited in depressions

▲ (Top) *The beautifully patterned and coloured Red-bellied Dace* (Phoxinus erythrogaster). (Above) *The golden morph of the Fathead Minnow* (Pimephales promelas) *is the main one encountered in the hobby.*

▼ *The Red Shiner* (Cyprinella lutrensis). *Older books on ponds and aquaria refer to this species as* Notropis lutrensis.

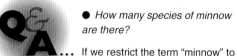

● *How many species of minnow are there?*

... If we restrict the term "minnow" to those species whose name actually contains the word "minnow", then there are around 30. If we expand it to include closely related "dace" species, the number rises to around 50, and if we also include "shiners" as well, the number rockets to around 165–170 species! However few of these, other than the ones referred to here, are likely to be available in large numbers commercially, either for ponds or coldwater aquaria, other than through specialist suppliers or members of specialist societies.

● *How can the different types of White Cloud Mountain Minnow be differentiated?*

There are two types of White Cloud Mountain Minnow in the wild: the variety from around Canton has white edges on the dorsal (back) and anal (belly) fins; that from around Hong Kong has red edges to these fins. Some people regard these two varieties as subspecies, while others claim that the Hong Kong type has developed from released/escaped aquarium specimens. There is also at least one other type: a cultivated long-finned variety that is best kept in aquaria.

main criterion when choosing fish for a pond or water garden, especially one with a wildlife emphasis. Consequently, even the wild-type Fathead Minnow came to be in demand for ponds, despite the fact that the dark dorsal colours of this and other species in the group made them less easy to see than cultivated colour morphs.

All the species outlined here are hardy, as long as conditions are not extreme. However, the long-finned variety of the White Cloud Mountain Minnow is less robust and will suffer under prolonged cold temperatures.

The minnows described here eat a wide variety of live and commercial foods. Yet bearing in mind their size, large hard foods, such as Koi/Goldfish pellets, should be ground or pre-soaked.

Miscellaneous Cyprinids

While the selection of species discussed in the preceding pages accounts for most of the cyprinids kept in ponds in temperate areas, there is

▲ *Dace* (Leuciscus leuciscus) *are close relatives of the Orfe. These fish are fast, active swimmers that feed on insects and crustaceans, but will also take dried foods.*

a whole host of other species that can be considered, depending on locality, personal preference and type of water scheme. Perhaps surprisingly, water gardeners in tropical countries rarely keep any native cyprinids in their ponds, preferring instead the more traditional types such as Goldfish and Koi for their size and colour. It is therefore predominantly temperate enthusiasts from northern Europe who favour wildlife schemes who have popularized these other cyprinids, most of which are drably coloured and some of which are not always in full view.

The accompanying table lists the most familiar species, and provides brief details on overall size and characteristics. For further information, the reader is referred to relevant wildlife reference sources.

● *Will wild species interbreed if kept together in the same pond?*

... Most species will not interbreed, while some won't breed at all, unless the pond is large enough and can provide adequate spawning conditions. Some species may, however, interbreed, either intentionally or by accident, if their spawning periods and habits coincide. The most likely species are: Crucian Carp, Gibel/Prussian Carp and Goldfish – which all belong to closely related species of the same genus, *Carassius*; and Rudd and Roach – which belong to separate genera, *Scardinius* and *Rutilus*, respectively.

● *Can I mix wild species with ornamental pond fish?*

If wild species are adequately quarantined and do not pose either a health threat or any other type of risk, e.g. predation, they can be kept with ornamental fish. I have, for example, kept a self-perpetuating population of Gudgeons (*Gobio gobio*) in the same pond as my ornamentals for many years without the slightest problem to either party. Wild species from hatcheries or specialized aquatic outlets will have been commercially bred or legally collected. If, however, you plan to collect your own stocks, it is essential to check that you are not contravening any laws and to collect (and transport) the fish with the absolute minimum of stress.

Some Miscellaneous Cyprinids

maximum adult size

Crucian Carp 20in (51cm)
Carassius carassius
Similar to Goldfish, but larger, much deeper-bodied, convex dorsal fin (in the Goldfish, it is straight or concave) and smaller head

Gibel, Prussian Carp 14in (36cm)
Carassius gibelius
"Western" version of the Goldfish (which is Asiatic in origin). Similar to Goldfish but deeper-bodied and with smaller head. Dorsal fin profile is straight. Colour: greenish-brown on the back and golden-green on the sides

Clicker Barb, Stone Moroko 4in (10cm)
Pseudorasbora parva
Slender and hardy. Makes clicking noises, particularly during courtship/aggressive displays

Bream 31in (79cm)
Abramis brama
Deep, flat-sided body with large, easily dislodged scales. Ventrally placed mouth which is extendible and, thus, adapted for bottom feeding

Bitterling 3.5in (9cm)
Rhodeus sericeus
Lays eggs inside water mussels. Asian species, such as *R. ocellatus*, *Tanakia tanago* and *Acheilognathus* spp. also occasionally available

▼ *Bitterling*

Bleak 6in (15cm)
Alburnus alburnus
Slender, silvery fish with flattened sides. Scales easily dislodged (as in the Moderlieschen). Large eyes. Surface shoaler

Moderlieschen, Opaline, Belica 4.8in (12cm)
Leucaspius delineatus
Slender, silvery species with easily dislodged scales. Surface shoaler

Dace 12in (30cm)
Leuciscus leuciscus
Slender, silvery fish with small mouth and pointed head. Spends the day in midwater or near the bottom, but rises to the surface towards evening

Gudgeon 8in (20cm)
Gobio gobio
Slender, bottom-dwelling species which will learn to feed from the surface of the pond

Silver Carp 39in (1m)
Hypophthalmichthys molitrix
Large head, compressed belly and "keel" along the ventral region. Upward-pointing mouth; small scales. Feeds exclusively on phytoplankton (free-floating microscopic plants) and therefore difficult to keep in peak condition in most ponds

Chub 24in (61cm)
Leuciscus cephalus
Slender and silvery; blunt head and large mouth

Pale Chub, Zacco 7in (18cm)
Zacco platypus
Attractively marked, Japanese species. Fast-swimming predator with excellent jumping ability. Males have unusually shaped anal fin

Other Temperate Fish

THE BORDERLINE BETWEEN WHAT CONSTITUTES a temperate fish and what constitutes a tropical one is hard to define. Indeed, it might be more accurate to see them as overlapping rather than as exclusive categories, since many so-called temperate fish can tolerate tropical conditions and vice versa. For our purposes, those miscellaneous species that are more clearly temperate are treated here, while those normally thought of as tropical are covered elsewhere (see "Coldwater" Tropical Fish, pages 108–115).

Bullhead Catfish

Family: Ictaluridae. These North American catfish are – both physically and behaviourally – very interesting, but they are also predatory.

● *Are there any fish that are unsuitable for keeping in ponds?*

... It is impossible to give an all-embracing answer to this question, because even species that are poorly suited for pond life can be kept if appropriate conditions can be provided. Generally speaking, though, fish that require cool, running, oxygen-rich water, like the Rainbow Trout (*Oncorhynchus mykiss*), cannot be catered for in most garden ponds. Other unsuitable fish are large, highly predatory species, such as the Northern Pike (*Esox lucius*) and its relatives, but, even here, if the pond is large enough, namely of lake proportions, contains abundant submerged vegetation and suitable prey, it is possible to keep one or a few Northern Pike or Pickerels.

● *How can bottom-dwelling fish be encouraged to come to the surface, and so be more frequently seen?*

Although certain species and varieties of pond fish are primarily midwater or bottom dwellers (this includes even Goldfish and Koi), few are so rigidly bound by biology to the lower levels that they will resist all

Some of the madtoms (*Noturus* spp.) are very small, e.g. the Pygmy Madtom (*N. stanauli*; 1.5in (4.1cm), while others, such as the Stonecat (*N. flavus*) can grow to around 12in (30cm). The bullheads, e.g. the Brown Bullhead (*Ameiurus nebulosus*), which grows to around 20in (51cm), are much larger, while the Channel Catfish (*Ictalurus punctatus*; the most widely available species, especially in its albino form), is larger still, attaining a length of around 50in (1.25m).

▼ *The Sterlet* (Acipenser ruthenus) *is a beautiful fish that is becoming very popular. It grows up to 4ft (1.2m), so make sure that you have room to accommodate it.*

attempts to lure them to the surface. Keeping such fish in the company of surface feeders like Orfe, or midwater/surface types like Rudd, will often bring positive results. Feeding floating foods also helps. In fact, with a little patience at feeding time, and so long as you ensure that no sudden movements frighten the fish, you can get most bottom dwellers to come to the surface on a regular basis within just a few weeks of their introduction to the pond. Once the fish feel safe, they will often leave the bottom, even outside feeding time, and may join their pondmates in midwater.

All of these catfish are hardy species that can be accommodated in temperate ponds. However, their predatory nature dictates what other fish can safely be kept with them.

Sturgeons

Family: Acipenseridae. Sturgeons can be giant-sized creatures, with species such as the Beluga (*Huso huso*) growing to a maximum length of over 16ft (5m). As far as garden ponds are concerned, though, only two species are available in any numbers: the Sterlet (*Acipenser ruthenus*) which grows to around 47in (1.2m) and the very attractively scaled Stellate or Star Sturgeon (*A. stellatus*), reaching 60in (1.5m). Even at these relatively modest sizes (in sturgeon terms), both species are large and can, obviously, only be kept in substantial ponds. Not surprisingly, therefore, sturgeons have begun to attract the attention of Koi keepers, whose sediment-free, clear-water set-ups can not only house large fish, but also allow dark-bodied, bottom-living species to be seen and appreciated, even in deep water.

Mudminnows

Family: Umbridae. Despite their name, Mudminnows are very different to the minnows discussed earlier. Mudminnows are predatory, and exhibit clear similarities to both pikes and pickerels (family: Esocidae). There are around five species in the family, of which two are occasionally available. The Eastern Mudminnow (*Umbra pygmaea*) is native to the eastern seaboard of North America, but has been introduced into many locations that lie outside its natural range, including several European countries. It grows to around 4.3in (11cm). The European Mudminnow (*U. krameri*) can grow slightly larger, to 5.1in (13cm).

Both are tough and resistant to a wide range of adverse conditions. This resilience, however, must not be used as a reason for complacency. Good water quality must be maintained at all times for this bottom-dwelling species.

Loaches

Family: Cobitidae. Several species of loach, none exceeding around 12–14in (30–35 cm) in length are occasionally on offer. Some, such as the European Weather or Pond Loach (*Misgurnus fossilis*) – which is reported to grow to 14in (35cm) in the wild – become hyperactive when there is a drop in barometric pressure, as happens prior to a storm. At such times it can be seen gulping air at the water surface, as can the closely related Dojo, Japanese or Chinese Weather Loach (*M. anguillicaudatus*), which can grow to 8in (20cm) and is also available in a golden form. The Stone Loach (*Noemacheilus barbatulus*), which attains a length of around 6in (15cm), and the Spined Loach (*Cobitis taenia*), reaching 4.7in (12cm), are two other European species that can sometimes be found for sale. All species like a soft substrate in which to bury themselves or root for food.

A more tropical species (which is hardy but cannot withstand extreme cold) is the so-called "Coldwater" Loach (*Botia superciliaris*), which is attractively marked and grows to 3.2in (8cm).

Sticklebacks

Family: Gasterosteidae. Two species of sticklebacks, the Three-spined (*Gasterosteus aculeatus*) and the Nine-spined (*Pungitius pungitius*) are fairly frequently available. Both are widely distributed in Europe and North America, with the Three-spined Stickleback exhibiting considerable variation in its few scales (called scutes) throughout the range. The Three-spined, at 4in (10cm) is larger than the Nine-spined, at 2.8in (7cm), and looks the sturdier of the two species. Both are hardy and build nests consisting of fibrous vegetation. Male Three-spined Sticklebacks develop bright red coloration around their throat and build their nests on the bottom, while Nine-spined males develop black colours and pale blue pelvic spines and build their nests 2.8–4in (7–10cm) off the bottom.

Both species are suitable for wildlife ponds, as long as they do not contain any large predatory species. The presence of predators can result in injury to both parties, as the stickleback is damaged by being sucked into a predator's mouth and the predator gets an unpleasant surprise from the tough dorsal, anal and pelvic spines of the stickleback.

Sunfish

Family: Centrarchidae. This family consists of around 30 species, known variously as sunfish, bass, perch (not including the European Perch *Perca fluviatilus*), crappies, the Pumpkinseed, Flier and Warmouth. Of these, several, such as the Pumpkinseed (*Lepomis gibbosus*), have become popular, first as coldwater aquarium fish and, subsequently, as candidates for wildlife ponds. The colours exhibited by some species are particularly beautiful when viewed from the side and this has contributed to their increased popularity. However, it is the breeding behaviour of a few types that make these fish particularly interesting. In the Bluegill (*Lepomis macrochirus*), subordinate males can take on

◀ *The Weather Loach* (Misgurnus fossilis). *As its name suggests, this fish is something of a "biological barometer".*

quasi-female coloration, so gaining access to the territory of a dominant male and – in the process – being able to mate without being detected. Such males are known as "satellites". Alternatively, subordinate males may dash in and mate with a female while the dominant male is engrossed in his own breeding activities. These males that dash in and out again are referred to as "sneakers". In the Pumpkinseed and a number of other species, eggs are laid in a depression that is prepared and defended by the male. This behaviour resembles a spawning strategy more commonly seen in tropical cichlids (family: Cichlidae).

Most sunfish species are hardy and able to withstand low temperatures. Even so, some may not survive the extended winter conditions encountered in some northern temperate regions. Sizes within the family vary from as little as 3.7in (9.5cm) in the Bluespotted Sunfish (*Enneacanthus gloriosus*) to as much as 38in (97cm) in the Largemouth Bass (*Micropterus salmoides*). However, most species that are available to buy for either ponds or coldwater aquaria are 6in (15cm) or less in total length and will accept a wide variety of foods.

Darters

Family: Percidae. Although they belong to the same family as the predatory European Perch, American Yellow Perch (*Perca flavescens*) and Ruffe (*Gymnocephalus cernuus*), the darters are small (4in/10cm) North American and Canadian bottom-dwelling fish, some of which are under threat in the wild.

The largest genus, *Etheostoma*, contains some spectacularly coloured species, such as the Emerald Darter (*E. baileyi* – 2.4in/6.2cm), the Cherry Darter (*E. etnieri* – 3in/7.7cm) and the rare Sharphead Darter (*E. acuticeps* – 3.3in/8.4cm).

Q&A

● *Is it ever possible to make pond fish tame?*

... With relatively few exceptions, pond fish, irrespective of type, do not become so tame that they will allow themselves to be stroked or picked up. The exceptions to this general rule are usually found among Koi and large Goldfish, but, even here, this behaviour is only encountered in specimens that have been afforded the considerable time and patient persistence required for them to become fully accustomed to a human presence by the pool. Such fish are capable of being completely relaxed, as they are aware that no threat of any kind exists. Under such circumstances, some fish will even allow themselves to be lifted gently out of the water. Few of us can, however, spare the time that such bonding demands. As a result, while our fish may become used to our presence around the water's edge, may even "beg" for food or accept it from our hands, and may distinguish our shape from that of, say, a heron or cat, they will generally retain a certain degree of "healthy suspicion" and will shoot off if they detect any signs that they associate with danger.

● *Why are darters not seen more often in ponds?*

Two of the principal reasons why darters are rarely sighted are that they are modestly sized and bottom-hugging. However, if they are accommodated in a multi-pond system where the individual ponds are joined together by shallow streams or artificial channels, darters will seek out these stretches of moving water, where they can then be observed and fully appreciated.

"Coldwater" Tropical Fish

MANY TEMPERATE AREAS EXPERIENCE SUMMER temperatures that fall within the tolerance levels of at least some fish species usually thought of as tropical. In others, even winter temperatures can come within the range of a few of the hardier types like the Medaka (*Oryzias latipes*) and Mosquito Fish (*Gambusia affinis* and *G. holbrooki*). Farther south, but still well outside the tropics (e.g. in Mediterranean regions), winter weather can often be sufficiently mild to make keeping other hardy species such as the Paradise Fish (*Macropodus opercularis*) and the Guppy (*Poecilia reticulata*) in outdoor ponds a distinct all-year-round possibility. In tropical regions, of course, no temperature limitations exist and virtually any native species can be kept outdoors.

The following is a small selection of some possible "coldwater" tropical fish that, climate permitting, could be regarded as interesting, attractive and unusual permanent or temporary/seasonal choices. Their general requirements are also included and these must be checked for suitability against prevailing conditions so that only appropriate selections are made. This is especially important in regions where these species can only be kept safely in ponds during the summer months and for regions that are prone to major temperature fluctuations. If in any doubt, the safest thing is to err on the side of caution and make an alternative choice.

Guppy

Poecilia reticulata (family: Poeciliidae). After the Mosquito Fish, the Guppy (or Millions Fish) is the most widely distributed of the livebearers. Moreover, since it occupies such diverse habitats as ditches, streams and river tributaries, some of which are brackish, it is hardly surprising to find that wild Guppies occur in a variety of size and colour combinations in the wild. Despite their differences, all wild Guppies have short fins.

Breeders and researchers have been exploiting the inherent genetic variability of the Guppy ever since its introduction to the European hobby in 1908, with the result that we now have endless strains that become progressively more elaborate with each passing year. Of the numerous varieties that exist, the fancier ones, which have long, flowing fins, are slower swimmers than their shorter-finned counterparts and may form easier targets for predators. Therefore, in areas where such predators exist, shorter-finned varieties would appear to constitute a better choice for ponds. Long-finned types are also generally less hardy than shorter-finned ones and will be more prone to succumb to environmental fluctuations, especially sharp drops in temperature.

Guppies, as all members of the family Poeciliidae (with one exception, *Tomeurus gracilis*) do not lay eggs. Instead, these are fertilized inside the female, where they remain within their egg

---

Factfile · *Guppy*

SIZE: Wild Guppies: up to 1.2in (3cm) for males and 2in (5cm) for females. Cultivated varieties are usually larger than this.

DISTRIBUTION: Widely distributed north of the Amazon basin in Antigua, Barbados, Dutch Antilles, Grenada, Guyana, Leeward Islands, St. Thomas, Trinidad, Venezuela and Windward Islands. It is also found in numerous exotic locations, including Mexico, some European countries, Australia and Singapore.

HABITAT: Slow-moving or standing bodies of water containing submerged vegetation are preferred. Sometimes found in brackish areas.

TEMPERATURE: c.15–28°C (59–82°F).

DIET: Wide range of commercial and live-foods. The diet should include a vegetable component (encrusting and free-floating algae found in most ponds will provide an adequate supply).

BREEDING: Depending on the size of the female, broods consisting of well over 100 fry can be produced every 4–6 weeks during the summer season. However, 30–40 fry per brood is more common.

sacs (follicles) until they hatch several weeks later. At that point, the female gives birth to fully formed young (known as fry), which are self-sufficient from the outset and should be offered protection from their parents in the form of fine-leaved vegetation.

The Guppy has been known scientifically as *Poecilia reticulata* since 1963. Before that, it had a variety of names, the most well-known of which was *Lebistes reticulatus* (still occasionally found in older aquatic literature). Its common name derives from *Girardinus guppii*, the name given to it in 1866 in honour of the Revd. Robert John Lechmere Guppy, who is sometimes re-ferred to as the discoverer of this fish in Trinidad. He is, however, more likely to have rediscovered it, since the species was first scientifically des-cribed in 1859 by Peters, who based his findings on specimens from Caracas in Venezuela.

▼ *Guppies* (Poecilia reticulata) *should not be kept with large Koi or Orfe, which may eat them, nor with Mosquito Fish, which harass them. To avoid fin-nipping, the shorter-finned varieties are better for ponds.*

● *Why is the Guppy also called the "Millions Fish"?*

... The origin of this (now almost defunct) name is uncertain. One explanation may be the Guppy's mass shoaling and wide distribution in the wild. Alternatively (and more likely), the term may refer to the diversity of wild male Guppies, which are all different in one way or other. Thus, a million male specimens display a million permutations on the basic shared colour theme.

● *Will Guppies survive temperate winters outdoors?*

Though a robust species, Guppies cannot tolerate the low winter temperatures of most temperate countries. Where established populations have been reported from such areas, they have invariably been associated with warm-water discharges, such as outflows from power stations or – as in the US states of Idaho and Wyoming – thermal springs. Elsewhere, while breeding populations are known to exist in non-tropical areas, (e.g. southern Europe), these are all in locations where water temperatures do not drop below c.10°C (50°F) for any length of time during winter. This species must therefore be brought indoors to a heated aquarium in temperate zones.

Mosquito Fish

Gambusia affinis, G. holbrooki (family: Poecili-idae). Mosquito Fish are small North American livebearing species. With their slick, torpedo-like body shape and voracious appetite for any tiny creature that is small enough to swallow (including their own new-born young), Mosquito Fish are predators *par excellence*. Their partiality for aquatic insects is such that, in the early 1900s, they were seen as a potentially devastating biological weapon in the fight to eradicate malaria from all tropical and subtropical regions of the world where this disease existed. As a result, Mosquito Fish were introduced into numerous countries outside their natural ranges, where, true to form, they consumed vast numbers of malarial mosquito larvae and pupae. (In this,

▼ *A trio of Mosquito Fish* (Gambusia affinis). *The front male has extended his gonopodium (modified anal fin) in preparation for mating with the (larger) female.*

Factfile *Mosquito Fish*

SIZE: Males: 1.2–1.6in (3.2–4cm); females: up to 2.8in (7cm).

DISTRIBUTION: *G. affinis*: Río Panuco basin, northern Veracruz in Mexico; northward to southern Indiana and east to Alabama, including the Mississippi drainage system and Texas. *G. holbrooki*: From Central Alabama, eastward to Florida and northward along the Atlantic drainage to New Jersey. Both species have become established in many exotic locations.

HABITAT: Slow-flowing or still waters with abundant submerged vegetation. May also occur in brackish habitats.

TEMPERATURE: c.10–30°C (50–86°F).

DIET: Predominantly carnivorous, consuming insects, crustaceans and small fish, but will accept virtually any other food offered.

BREEDING: Broods of 10–80 fry produced every 5–8 weeks during spring and summer.

they were even more efficient than Guppies, which were the first fish tried for malarial control.) Although these introductions were well-intentioned, they were nevertheless somewhat misguided. Rather than securing a biological victory over malaria, they have resulted in unprecedented population explosions of Mosquito Fish in areas of the world where they should never have been present in the first place. A further unforeseen negative consequence in most such regions has been that the native fauna has suffered, sometimes to the point of being driven into local extinction.

Despite their notorious reputation, Mosquito Fish are exquisite little fish adored by many hobbyists, particularly livebearer enthusiasts. Traditionally, two very similar fish have been regarded as the "true" Mosquito Fish. Up to 1988, these were referred to as *Gambusia affinis affinis* (the Western Mosquito Fish) and *Gambusia affinis holbrooki* (the Eastern Mosquito Fish), i.e. two subspecies of *Gambusia affinis*. Today, both these fish are regarded as full species in their own right, and have been reclassified as *Gambusia affinis* and *G. holbrooki*. In the latter species, the males have a tendency towards melanism (i.e. having either black spots or a totally black body).

Other Livebearing Species

Family: Poeciliidae. There are several other species of livebearers that are popular among tropical aquarists and which are sometimes transferred to outdoor ponds during the summer season in temperate areas. Those that may be considered include: Swordtail (*Xiphophorus helleri*), Southern Platy or Moonfish (*X. maculatus*), Variatus or Sunset Platy (*X. variatus*), Sphenops/Black (Short-finned) Molly (*Poecilia sphenops*), Sailfin Molly (*P. latipinna*) and the Sailfin/Yucatán Molly (*P. velifera*). In regions experiencing warmer winter conditions, where water temperatures do not drop below 15–18°C (59–64°F), it is possible that some short-finned swordtails and platies may survive from one season to the next in deep ponds. However, none of these species can be considered as hardy as either the Guppy or the Mosquito Fish, although all have become established as breeding populations outside their natural range.

All these species can be obtained in a wide range of colour and fin variations, with the shorter-finned types being hardier than the long-finned ones. Since they are of the same genus, all three *Xiphophorus* species can interbreed with one another and produce viable hybrids. The same is true of the three Molly species.

Q&A

● *Are there any other Mosquito Fish?*

... At least one other livebearer, *Heterandria formosa*, is also referred to as the Mosquito Fish. It is a much smaller fish than either *Gambusia affinis* or *G. holbrooki* and does not possess their keen predatory instincts. *Heterandria formosa* (also known as the Dwarf Livebearer or Dwarf Topminnow) is a rather timid fish which should not, under any circumstances, be kept with either of the two other Mosquito species.

● *Can Mosquito Fish be kept with other species?*

It is possible to keep Mosquito Fish with other species, but only if these are hardy and resilient and if they do not possess long flowing fins or otherwise present "desirable" targets. Goldfish, for example, irrespective of variety, and Mosquito Fish should be kept well apart. The same applies to Guppies and, indeed, most other smallish species.

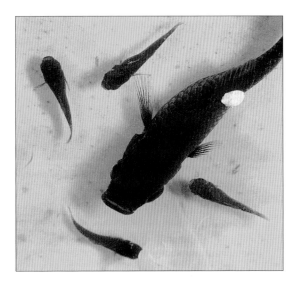

▲ *A Black Molly* (Poecilia sphenops) *and its young pictured in a shallow pool. These fish are tolerant of brackish and even marine conditions.*

Japanese Medaka

Oryzias spp. (family: Oryziidae/Oryziatidae). Several species of this genus are available, usually for tropical and coldwater aquaria. However, the Japanese Medaka, Ricefish or Geisha Girl (*O. latipes*) is a hardy species that is also suitable for many temperate ponds. It has been known to survive mild winters in deep ponds, even as far north as the southern counties of England, but caution is advised, owing to the unpredictable severity of winters in such latitudes.

The wild type of the Japanese Medaka is only rarely found in shops these days, its place having been taken almost entirely by the commercially developed golden morph. Indeed, the original common name is rapidly being supplanted by "Golden Medaka". These fish are active, peaceful shoalers that prefer the upper layers of the water column. They are therefore always in full view during the warmer months of the year, making them an ideal choice for most ponds in summer. As winter approaches, they tend to congregate in deeper waters and, during the very coldest periods (as long as these fall within their

temperature range), will hover just above the bottom of the pond, but will rise to the surface again on warm days.

Other species of ricefish include: *O. celebensis* (Celebes Medaka); *O. javanicus* (Javanese Medaka), *O. melastigma* (Spotted Medaka) and the rarely seen *O. nigrimas* (Black Medaka). None, however, are as hardy as *O. latipes*.

Dwarf Sunfish

Elassoma spp. (family: Elassomatidae). At present, there are fewer than 10 recognized species of *Elassoma* (most references only quote six). Of these, the one most often encountered is the Everglades Pygmy Sunfish (*E. evergladei*), more frequently on offer as an aquarium fish, rather than for ponds. It is also known as the Everglades Dwarf Sunfish, or Florida Pygmy Sunfish.

Although *Elassoma* species may well survive in many temperate-water pond systems, there are two main factors which need to be considered: Firstly, owing to their small size and relatively dark coloration, they can easily "get lost" in a large pond. Also, unlike some other small fish, such as the Medaka, dwarf sunfish are bottom dwellers that rarely rise to the water surface and are, consequently, only occasionally seen, other than in shallow ponds. Whether or not they can be regarded as "good" pond fish therefore probably depends – assuming that environmental conditions are within their limits of tolerance – more on the pond keeper's/water gardener's personal criteria regarding suitability, than on the ability of the fish to survive.

Some classifications group the dwarf sunfish with their larger namesakes within the family Centrarchidae. However, dwarf or pygmy sunfish possess several features that separate them from the centrarchid sunfish. The most obvious of these are the total absence of a lateral line (the line-like organ consisting of a row of sensory pits that many fish, including sunfish, exhibit on the side of the body) and the possession of a rounded caudal fin, rather than a forked one. In addition, as their name implies, dwarf sunfish are far smaller than their centrarchid cousins, the largest species – the Banded Pygmy Sunfish (*Elassoma zonatum*) – only growing to a length of 1.9in (4.7cm).

Factfile   *Japanese Medaka*

SIZE: Up to 1.6in (4cm).

DISTRIBUTION: China, Japan and South Korea, but has also been reported from Java and Malaysia.

HABITAT: Medakas occur in a variety of waters in the wild, preferring small streams, and, as one of their common names indicates, rice paddies.

TEMPERATURE: c.15–28°C (59–82°F).

DIET: A wide range of livefoods and commercial diets is accepted.

BREEDING: One of the ricefish's most unusual characteristics, which is common to all species, is that the females will carry their egg clusters attached to their anal/genital aperture for a while, until they deposit them, usually in clumps, among fine-leaved vegetation. These egg clusters resemble a tiny bunch of grapes. Another unusual feature is that the eggs may be fertilized externally (as in true egglayers) or internally (as in livebearers). Hatching takes about 7–10 days, depending on the temperature of the water.

Factfile — *Everglades Dwarf Sunfish*

SIZE: Around 1.4in (3.6cm).

DISTRIBUTION: Along the Atlantic and Gulf Coastal Plain drainages of North America, stretching from North Carolina, through Alabama, to Florida, as far south as the edge of the Everglades.

HABITAT: Heavily vegetated still or slow-moving waters.

TEMPERATURE: c.10–30°C (50–86°F).

DIET: Wide range of small foods accepted, particularly livefoods.

BREEDING: Eggs are scattered among fine-leaved vegetation and hatch 3–4 days later, depending on temperature.

Q&A...

● *At what stage should Golden Medakas be removed from an outdoor pond in temperate areas?*

This is hard to specify. Although Medakas usually live in the range 15–28°C (59–82°F), they have been kept in ponds where surface water temperatures fell far below this minimum. However, as these ponds were deep (24in/60cm or more), the temperature near the bottom, where most fish gather in winter, would have been slightly warmer than at the surface. Medakas have even survived a winter in southern England, and bred during the following season, though this is exceptional. So, to be safe, bring Medakas indoors when outdoor surface water temperatures begin to drop below 15°C.

● *Are Golden Medakas tolerant of each other?*

Medakas are generally tolerant, and should be kept in a shoal. Nevertheless, a hierarchy develops among males – particularly in the breeding season – and, if only two males are present, the subordinate one will usually be driven away from the group. The overall condition of such lone males may deteriorate over time. It is therefore best to keep Medakas either in a small group consisting of a single male and several females, or (preferably) as a larger group of some 10–12 fish, including at least three or four males.

● *How temperature-hardy are dwarf sunfish?*

Dwarf sunfish are normally very hardy. For instance, *Elassoma evergladei* can survive (aquarium) winter temperatures as low as 8–10°C (46–50°F) without any ill effects.

American Flagfish

Jordanella floridae (family: Cyprinodontidae). This striking fish is one of a group of tropical and semitropical egglaying cyprinodonts collectively known as "killifish". The American Flagfish derives its common name from the distinct horizontal red markings on its body and its blue head, which resemble the "Stars and Stripes".

The American Flagfish was first introduced into the European fishkeeping hobby around 1914–15 as a coldwater species, and was subsequently "adopted" by the tropical side of the hobby. It is now so strongly identified with the tropical aquarium that many new aquarists are surprised to learn of its coldwater origins.

Although the American Flagfish is a hardy species that thrives in relatively cool waters, it must be stressed that it cannot be considered as a permanent pond resident in temperate zones. There are no known instances of American Flags surviving typical temperate winters in outdoor ponds. However, they make good temporary pond residents, and can be kept outdoors for five months, or even a little longer.

Factfile — *American Flagfish*

SIZE: Up to 2.6in (6.5cm) for males; females smaller.

DISTRIBUTION: From Florida (USA) southwards to the Yucatán in Mexico.

HABITAT: Slow-flowing or standing waters with ample submerged vegetation. May also be found in brackish conditions.

TEMPERATURE: c.16–25°C (61–77°F).

DIET: Omnivorous; a wide variety of foods will be accepted. A regular vegetable component is required (most ponds will provide adequate supplies of this dietary component). Algae and soft-leaved plants are especially favoured.

BREEDING: American Flagfish exhibit two different breeding strategies: they either scatter their eggs among fine-leaved vegetation or else lay them in a male-excavated and male-defended depression on the bottom. Hatching (depending on temperature) takes 6–9 days. The male continues to care for the fry until they are large enough to fend for themselves.

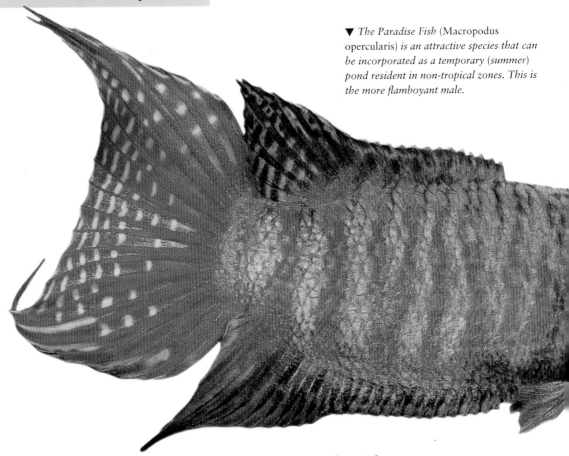

▼ *The Paradise Fish* (Macropodus opercularis) *is an attractive species that can be incorporated as a temporary (summer) pond resident in non-tropical zones. This is the more flamboyant male.*

Factfile — *Paradise Fish*

SIZE: Up to 4.7in (12cm) for males, but usually smaller.

DISTRIBUTION: Southern China, Korea, Taiwan (and some surrounding islands) and Vietnam.

HABITAT: Shallow, still or slow-flowing waters with aquatic vegetation, e.g. rice paddies.

TEMPERATURE: 16–28°C (61–82°F).

DIET: Wide range of foods accepted.

BREEDING: Males build bubble-nests at the water surface, usually under a floating leaf, below which spawning embraces take place. The eggs, which take about 1 day to hatch, are blown into the bubbles by the male, who stands guard over them until they hatch. Females have occasionally been known to assist in these tasks, but usually don't. Young fish should be brought indoors and kept in tanks well before the temperature drops close to the species' lower tolerance range.

Paradise Fish

Macropodus opercularis (family: Belontiidae). This species of Paradise Fish was the first species of tropical fish to be imported into Europe, in 1869, since when it has been popular with aquarists, owing to its beautiful colours, interesting breeding behaviour and – most relevant from our point of view – its temperature tolerance. Although comfortable at temperatures as low as 16°C (61°F), it can survive in even colder water, provided the temperature drop is gradual. This applies more to the wild type of the species than to some of the colour varieties developed over the years, the least hardy of which appears to be the albino.

Other "Coldwater" Tropicals

In addition to the species profiled above, there are many others which, while being classified as tropical, can nevertheless tolerate temperatures that are below the recognized tropical range, i.e. below around 22°C (72°F). These "coldwater"

● *Can Paradise Fish be kept in shoals?*

... Although females will often associate with each other, adult male Paradise Fish tend to become aggressive, both towards other males and, to a lesser extent, towards females. It is therefore best to keep just a single male and several females, if the pond is small (i.e. length under 6.5ft/2m). Larger ponds can generally house more males, especially if they offer easily defended territories and adequate stocks of submerged vegetation.

● *Can Paradise Fish be kept with other fish?*

Most species of pond fish that are a similar size are usually tolerated by male Paradise Fish, although the situation should be constantly monitored, as this cannot be regarded as a totally "risk-free" species. For their part, mature male Paradise Fish can become the focus of unwelcome attention from fin-nipping species such as Mosquito Fish, which find their extended fin rays an irresistable temptation.

● *Are American Flagfish peaceful towards other fish?*

Certainly, *Jordanella* is tolerant of other species, but males do tend to be aggressive towards each other. However, since this species is not as strong a shoaler as, say, the Golden Medaka, lone subordinate males do not appear to suffer to the same extent. This is especially the case in a pond environment which offers hiding places and easily defended territories during the spawning season, when aggression is at its height.

● *Is the Sailfin Sucker a peaceful fish?*

Despite its large size, this is a generally peaceful, though powerful, fish. Some keepers have reported it to be timid and to be more active during the evening. However, many specimens I have seen have been active during the day and have mixed well with other sizable pondmates. This is a species about which we still lack essential details, and whose sheer size might pose a threat to small fish – it can grow to around 24in (60cm) in length.

● *How hardy is the Sailfin Sucker?*

Winter temperatures in its native China – especially in the north – fall below 15°C (59°F), making the species suitable for cool-water environments. However, whether or not it can withstand long periods at low temperatures in temperate-area ponds is, as yet, not known for sure.

tropicals can therefore be housed for varying lengths of time and with varying degrees of success in some non-tropical environments during the warmest months. During this time, such species can develop excellent coloration and may even breed, but a close watch on the temperature is essential so that the fish can be moved indoors at the first sign of colder weather.

Some fish species that qualify as "coldwater" tropicals include: the Chinese Sailfin Sucker (*Myxocyprinus asiaticus*), the Bronze Catfish (*Corydoras aeneus*), the Peppered Catfish (*C. paleatus*), the Rosy Barb (*Barbus conchonius*), the Green/Half-striped/Golden Barb (*B. semifasciolatus*), the Butterfly Goodeid (*Ameca splendens*), the Orange-tailed Goodeid (*Xenotoca eiseni*), the Blind Cave/Mexican Tetra (*Astyanax mexicanus*), and the Blue Acara (*Aequidens pulcher*). Detailed requirements of all such species should be sought in standard tropical aquarium books before a decision to introduce them into outdoor ponds is taken.

▲ *The Chinese Sailfin Sucker* (Myxocyprinus asiaticus) *is usually offered for sale as a small (4in/10cm) specimen. Eventually, though, it grows to a very substantial fish.*

Pond Plants

A POND THAT IS DEVOID OF PLANTS LOOKS
incomplete. Happily, there is a growing trend
towards regarding ponds as integral components of
water garden schemes, rather than as fish-only reserves.
Even Koi pools, whose voraciously herbivorous inhabitants
might seem to preclude the keeping of plants, are beginning
to reflect this new way of thinking. A wise choice of plants
and appropriate planting techniques can result in a degree of
compatibility that was once thought unattainable.

Plants come in many shapes, sizes and colours and have
traditionally been grouped into four main categories: submerged,
floating, marginal and surface plants. In recent years, some pond-
keeping authorities have distinguished bog plants from marginals
and regarded them as a new, fifth category. While this book recognizes
that there may be a subtle botanical distinction, it prefers to treat
bog plants in the practical context in which pond keepers encounter
them – namely, as part of a spectrum of marginal planting options,
which also includes shallow-water and deep-water marginals.

Another recent trend has been to see the water garden as a specialized,
but nevertheless integral, part of the total garden scheme. As a result,
many water garden centres now stock a comprehensive selection of
terrestrial plants for the areas immediately surrounding ponds.

Clearly, coverage of such "normal" garden plants is beyond
the scope of this book. The pages that follow will therefore
focus on the true pond plants, beginning with those that
are suitable for the moist areas around the pond edges,
and proceeding by stages to those that live
completely submerged.

▶ *A wide range of different types of pond plant has
been used to create this striking water garden.*

Why are Plants Important?

IT IS POSSIBLE TO APPROACH THE SUBJECT OF pond plants from two ostensibly very different perspectives. From a purely human standpoint, it is probably true to say that pond plants are included in a water scheme mainly for their aesthetic appeal. Without decrying this viewpoint (after all, every pond keeper buys certain plants for their sheer beauty), there is another, broader perspective that warrants serious attention. This places the needs of the pond and its inhabitants at the top of the agenda, and makes them the main criteria in choosing plants. Neglecting this aspect can result in an unbalanced, and sometimes unmanageable, water garden.

The Significance of Photosynthesis

The description of rainforests as "the lungs of the world" is no exaggeration. As air-breathing creatures, we humans, along with all other animals and plant species, need oxygen to survive. But where does it come from? Neither we nor the vast majority of other living organisms are

▲ *Filamentous green algae – or blanket weed – buoyed up on oxygen bubbles generated by photosynthesis.*

▶ *Pond plants provide fish, such as these Koi, with places of shade and shelter.*

PHOTOSYNTHESIS IN POND PLANTS

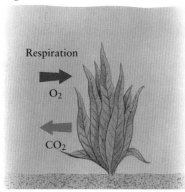

Day   Solar energy   Night

Respiration   Photosynthesis   Respiration

O_2 O_2 O_2

CO_2 CO_2 CO_2

◀ *During the day, photosynthesis outstrips respiration, with the net effect that carbon dioxide is taken in and a surplus of oxygen is generated. At night, as photosynthesis ceases in the absence of sunlight, no oxygen is generated. Yet respiration continues, resulting in a net intake of oxygen and an output of carbon dioxide.*

capable of generating oxygen. The only sources of this life-giving gas are green plants, which contain a complex green compound known as chlorophyll (responsible for their characteristic pigmentation). When subjected to light and provided with carbon dioxide and water, chlorophyll can combine these last two substances into food molecules (carbohydrates). Interestingly – and of vital importance for us – one of the by-products of the long and sophisticated series of reactions required to synthesize carbohydrates is oxygen. The whole process whereby carbon dioxide is used up, food is produced and oxygen is generated is called photosynthesis.

Plant Activity at Night

Since photosynthesis requires light, it can only occur during daylight hours. Once night falls, it stops completely. At the same time, the process of respiration, which requires oxygen and produces carbon dioxide (and which occurs, both day and night, even while photosynthesis is taking place) continues unabated. In net terms, green plants produce more oxygen during the day, through photosynthesis, than they need for respiratory purposes. The surplus passes out into the surrounding atmosphere (or water), so providing all the other organisms that need oxygen with this essential gas. At night the situation changes; once photosynthesis ceases, no more oxygen is generated, and so, in net terms, green plants take in oxygen from their surroundings to meet their respiratory requirements at this time.

Other Roles of Plants

In addition to oxygenating via photosynthesis (at least, in submerged species), pond plants perform a number of other valuable functions.

All types of pond plant possess roots through which they absorb a wide range of compounds from the water. Nitrates, for instance – which are among the end products of the natural detoxification of waste material in the pond (see Managing Water Quality, pages 182–183) – can rise to unacceptable levels if left unchecked. At excessively high concentrations, these may not only affect the health of delicate fish, but also act as excellent fertilizers for the nuisance algae variety known as blanket weed, or for those types that cause "green water".

By taking up some of these excess nitrates through their roots as fertilizer, pond plants perform the very valuable role of controlling algae biologically. Moreover, by absorbing other substances, they act as a cushion or "buffer", which, in turn, prevents any violent fluctuations in water quality.

Surface and floating plants (which may cover as much as 60% of the water surface) also create shade, shelter fish from bright light and predators, prevent evaporation from the pond, help cool the water, and act as food for some fish species or for the numerous small organisms that fish feed on. Many types of plant also offer fish a variety of spawning sites and provide a natural haven where eggs, fry and other forms of pond life can develop in safety.

Reproduction and Propagation

GIVEN FAVOURABLE CONDITIONS, POND PLANTS will often grow too large or reproduce too profusely for the space that you have set aside for them. This can occasionally happen in the course of a single growing season, with the spread becoming so vigorous that, unless appropriate steps are taken, a water scheme that was once attractive and well-balanced can soon end up looking a complete mess. In such circumstances, thinning is more appropriate than propagation in order to restore some form of balance. However, water gardeners will always relish the pleasure and challenge of propagating some of their own plants. Home-produced plants also make excellent gifts!

▲ *A bee on a water lily* (Nymphaea 'Evelyn Raudig'). *Pollination is an essential part of sexual reproduction in plants, and is most commonly effected by insects.*

Types of Reproduction

Plants reproduce in two ways: sexually and asexually (or "vegetatively").

Sexual Reproduction. Flowering in plants plays a far more important role than merely being aesthetically pleasing; it optimizes their chances of survival. The actual way in which a bloom is produced is a complex process that is still not fully understood. Flowering would appear to be triggered by responses to light and dark in the plants' leaves, via a pigment called phytochrome.

Subsequently, it is thought that a hormone is produced in the leaves and that this stimulates the production of blooms. Other factors, including temperature and certain hormones collectively referred to as gibberellins, may also be involved.

Every process undergone by living things performs some function or other in their survival. In the case of flowers, they contain the sexual organs of the plants that bear them and, within them, the "insurance ticket" for their continued existence. As in animals, the sexual organs of plants contain the female parts (pistils) and/or the male ones (stamens) which, in turn, contain the sex cells (gametes), i.e. eggs (ovules) or sperm (pollen). Within these sex cells are the nuclei and, inside them, the chromosomes that contain all the genetic information required to produce another plant of the same species.

During fertilization, which follows pollination, the genetic content of a male nucleus fuses with that of a female nucleus, so ensuring that the resulting fertilized cell contains equal male and female contributions. However, such fusion would result in a fertilized cell with twice the normal genetic material – unless there were a way of halving both the male and female parental contributions prior to fertilization. Indeed, this is precisely what happens. The process – known as meiosis or reduction division – creates male and female cells that contain half the normal genetic material. In other words, they have one gene of each kind instead of the normal two. Therefore, at fertilization, each sex cell provides half of the total genetic material needed, resulting in a fertilized cell that now contains the full complement of genes for the species in question.

What purpose does this halving and doubling serve? During the reduction process, genes are "moved about" in a process known as recombination, and mutations may occur. As a result, the new plant, while retaining all the characteristics of its species, will also hold new gene combinations which, over generations, give the species a degree of "adaptive flexibility" that asexually reproducing organisms cannot match.

Asexual/Vegetative Reproduction. Reproduction without the development or involvement of sexual organs such as flowers is said to be asexual or vegetative. The structures that allow plants to

Q&A

● *Is it difficult to propagate pond plants?*

... In many cases, propagating pond plants is considerably easier than propagating "normal" garden plants. This is especially so when using methods of vegetative propagation. One great advantage that aquatic and marginal plants have over their terrestrial counterparts is that, by their very nature, their roots (or the whole plant, in many instances) are immersed in water all the time, thus eliminating the risk of desiccation. Despite this, newly planted cuttings or offsets may require some "aerial" protection against strong sun or winds if they are placed in a location which is subjected to different environmental conditions than the parent plant from which they came. Propagating pond plants by seeds is either just as easy as for terrestrial species, or, in a few cases (e.g. water lilies), a little more challenging.

● *When is the best time to thin out or divide pond plants?*

Thinning and division of those species that grow vigorously should be conducted when they are in full growth, during spring and summer. This allows any surfaces that have been cut to heal quickly, whereas in lower water temperatures there is the risk of rot. Some bog plants, such as Hostas or Globe Flowers (*Trollius* spp.) can be divided in autumn.

▲ *The* Water Soldier (Stratiotes aloides) *is one of a large number of pond plants that can reproduce either sexually or, as in this case, asexually (vegetatively) through offsets.*

do this vary hugely in shape and size, despite their common function. Sometimes, as in Duckweed (*Lemna* spp.) and Fairy Moss (*Azolla* spp.), the only things initially visible are tiny plantlets/leaves/fronds growing out of the parent. These offspring then split off and grow independently until they, too, are mature enough to bud off their own offspring. Other plants, for example Arrowheads (*Sagittaria* spp.) produce elongated underwater stems that gradually develop swollen tips. These eventually fatten out into corms that split from the parent plant and float away to establish themselves elsewhere.

There are also rhizomes, branching modified stems that spread out from a parent plant at soil level or below (as in Irises; *Iris* spp.), and runners, stems that grow outwards (usually above soil level), laying down roots at certain points, from which new plants grow (as in the Water Fringe; *Nymphoides peltata*). Some other vegetative structures, such as stolons and bulbs, are not represented among pond plants.

Perhaps the main advantage of vegetative reproduction is that it allows a plant to establish itself very rapidly – sometimes in less than a season – at a particular location, if conditions are favourable. Indeed, certain vegetatively reproducing species can be extremely invasive.

The greatest disadvantage of vegetative reproduction is that, because it does not involve the "importation" of new genes from another plant or gene combinations from the same plant (as happens via flowers during sexual reproduction), the offspring are all identical genetic replicas of their parents – leaving aside the very occasional mutation. Therefore, while vegetatively reproducing plants may enjoy great success under one set of environmental conditions, this does not guarantee their survival under different circumstances. This is where the "survival insurance ticket" provided by flowers pays dividends. Yet most invasive pond plants have the best of both worlds, in that they also produce flowers.

Propagating Pond Plants

Pond plants may be propagated, either by sowing the seeds that result from the sexual reproductive activities of the species and/or varieties concerned, or vegetatively, by planting the various asexual reproductive organs/structures produced. Vegetative propagation is the only option for sterile hybrids or varieties (e.g. double types).
Seeds. When collecting seeds, you must wait until the fruits have ripened fully. Inspect daily as harvesting time approaches to ensure that the seeds are not dispersed by natural means.

Most seeds can be sown in standard seed trays containing soaked good-quality soil or special aquatic compost. Spread the seeds thinly over the surface and cover with soil, compost or sand. Lightly water marginal or bog plants, and for other types immerse the tray in a dish with

PLANTING A WATER LILY EYE

① ② ③

▲ *Water lilies produce "eyes" (growing points, from which shoots arise) along the rootstock (rhizome). To propagate the plant, cut an individual eye from the rootstock (1). Pot the cut sections of plant individually in good soil or aquatic compost (2), after dusting the cut surfaces with charcoal to prevent infection. Stand the pots in a tray of water (3) with the eyes just below the surface.*

DIVIDING AN IRIS RHIZOME

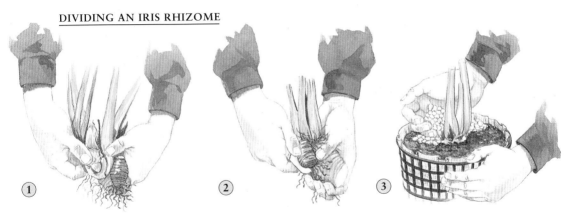

① **②** **③**

▲ *To propagate an Iris, you must divide the rootstock, which is also known as a rhizome. First lift and rinse off the plant, and then split the rhizome into sections with your hands (1). Ensure each section has several roots* *attached. Trim leaves and roots neatly with a knife (2), and discard any sections that show no new shoot growth. Replant carefully (3) without compressing the roots, and top-dress with small pebbles.*

Q&A...

● *Does pond plant propagation require very sophisticated equipment?*

No. Great success can be achieved using the following few items of equipment: seed trays; good compost or, preferably, special aquatic compost; hormonal rooting powder; planting baskets and lining; pebbles (for laying on top of compost after re-planting); frost-free facilities, e.g. greenhouse or garden frame; sharp knife.

● *What type of propagation produces the best results?*

If "best" means "quickest", in terms of ending up with a respectable display in a short time, then clump division as growth is beginning, or during the growing season itself, is about the best way of achieving good results. In terms of satisfaction or achievement, however, propagation via seed is the most fulfilling and therefore the "best" method, though it takes a long time.

● *Why can't double varieties and cultivars be propagated from seed?*

Double varieties have been developed from basic, single types. In so doing, every part of the flower undergoes changes, including the sexual organs themselves. This usually results in some degree (or even total) breakdown of the processes that are responsible for producing pollen and ovules in normally-bloomed plants and hence in the loss of gamete (sex cell) viability. In the case of cultivars, many can, indeed, be propagated by seeds, but the genetic make-up of the parents is often so mixed that such sowings frequently fail to grow true to type.

enough water to keep the soil thoroughly damp, or to cover the soil surface. As marginal plants begin to germinate, they need to be kept moist, while other types are best kept slightly covered with water. At the two/three-leaf stage, the tiny plants may be pricked out (as with "normal" garden plants) and cultivated until they are ready for transferring to the pond or water garden.

All these operations must be carried out in frost-free conditions (a garden frame or greenhouse is thus essential in frost-prone areas). **Vegetative Propagation.** The many strategies for propagating plants vegetatively are too numerous to list here. Rather, some general guidelines are given, which you can adapt to individual circumstances. In the case of rooted submerged species and many of the stemmed/leaved marginals, soft-stem cuttings offer several possibilities.

Cuttings of submerged plants need only be replanted in a basket or tray and kept under water until they strike roots and can be transferred to their final container, or the pond substrate. Weighted bunches of unrooted cuttings dropped into the pond will also root of their own accord. Cuttings from marginals should be trimmed of their lower leaves, dipped in a rooting powder, potted and kept moist until roots develop. Plants that form clumps are best propagated by dividing the clumps and replanting the individual portions. To propagate rhizomatous plants, see the accompanying diagrams.

Planting Your Pond

WHETHER THEY ARE SUBMERGED, SURFACE OR marginal, pond plants should ideally be grown in containers, though some wildlife and informal arrangements allow them to spread unrestrictedly. Keeping plants in containers makes them easy to control and in most cases does not adversely affect their growth or ability to flower.

Containers come in a wide variety of shapes and sizes, from small and medium-sized baskets to large crates. Both types have perforated sides, while some also have perforated bases. Large containers with a wide mesh must be lined with hessian or a similar material to stop the soil spilling out. However, smaller, micro-meshed baskets retain the soil considerably more effectively.

Solid plastic pots do not allow free circulation of water around the roots, and should only be used for those plants that take up most nutrients through their leaves (e.g. the majority of submerged oxygenators). Gross feeders, for example lilies and most marginals, should always be planted in lined, meshed containers that allow freer circulation of water and chemicals around the roots.

Nature versus Pond

In Nature, most pond and lake plants (as opposed to flowing-water types) have their roots anchored in a fine, organically rich, muddy bottom medium. Below the surface, the oxygen concentration is often very low and can contain toxic gases. Because of the large volumes of water in such ponds or lakes, these usually present no difficulties. In garden ponds, though, where water volumes are lower and fish concentrations higher, such toxic gases can be a problem. During the growing season, any bubbles of gas that escape from compacted and oxygen-starved rooting media will simply float to the water surface, burst

*Some **Marginal plants** can lend height to the water garden scheme.*

Water lilies give shelter and shade to fish and help keep the water cool.

Marginals by the edge will soften the harsh lines of the paving as they grow.

▲ *Planting suggestions for a vibrant and varied display in and around an informal pond. In reality, of course, all the plants shown will not flower at exactly the same time; if you choose plants carefully, you can ensure colour in your water garden from early spring to late summer. But even at other times, the different shapes and textures of foliage are extremely attractive.*

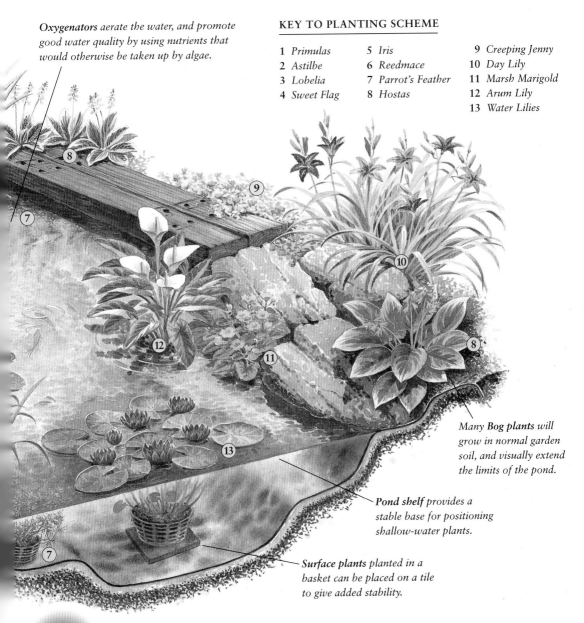

Oxygenators *aerate the water, and promote good water quality by using nutrients that would otherwise be taken up by algae.*

KEY TO PLANTING SCHEME

1 *Primulas*	5 *Iris*	9 *Creeping Jenny*
2 *Astilbe*	6 *Reedmace*	10 *Day Lily*
3 *Lobelia*	7 *Parrot's Feather*	11 *Marsh Marigold*
4 *Sweet Flag*	8 *Hostas*	12 *Arum Lily*
		13 *Water Lilies*

Many Bog plants *will grow in normal garden soil, and visually extend the limits of the pond.*

Pond shelf *provides a stable base for positioning shallow-water plants.*

Surface plants *planted in a basket can be placed on a tile to give added stability.*

Q&A...

● *Should a newly planted surface plant be located in its final position straight away or in stages?*

Some gardeners recommend that new surface plants initially be raised from the pond bottom, say on bricks, so that the top is near the water surface. In this method, established leaves on the plant look far better resting on the surface than suspended at an unnatural angle underwater. As new growth appears, the basket is gradually lowered. Others maintain that they should be put in their permanent, deep position from the outset; any fully unfurled leaves are simply removed before planting. Both methods work well, though the second has the advantage of being far simpler.

● *What shape or size of container should I buy?*

Square containers fit well in corners or along straight-edged shelves, while kidney-shaped ones are very good for informal ponds with irregular edges. Key points to look out for in all containers are stability and enough growing space for at least two seasons.

and disperse without any ill effects on the fish or other forms of free-swimming animal life. In winter, though, things can be very different if the surface of the pond freezes over. At such times, gas bubbles become trapped under the ice and dissolve in the water. It is thus wise to container-ize plants to ensure water circulation around their roots. Where containers have not been used, gently disturb the pond substrate regularly (but especially before winter) to disperse toxins.

Planting Different Types of Plants

Surface plants, especially water lilies, may be sold with bare roots. Check them thoroughly; if they have at least several healthy-looking shoots, few or no seriously damaged roots and a firm rhizome (or equivalent structure), they are a good buy. Old roots may seem darker and/or rougher than newer ones, but this is not a sign of poor health. If they are spongy or bruised, though, check that this damage does not extend back into the main rhizome.

Oxygenators are usually sold in bunches held together with lead clasps. It is safe to leave these on; although lead is toxic, the amounts involved are negligible. The clasps also keep the plants in place. Check bunched plants for any stem damage. In pot-grown oxygenators, look for healthy root growth through the rockwool medium.

● *Will pond plants grow well if they are not planted in baskets?*

Q&A ... For a submerged plant like Hornwort (*Ceratophyllum demersum*), which produces no roots, but grows as a mass of underwater stems and leaves, no planting medium at all is required. However, most submerged, surface and marginal plants do need a rooting medium. Whether this is put in planting baskets or spread out over the bottom of the pond is a matter of preference; the plants simply need enough to anchor their roots firmly. Baskets have the advantage of keeping the medium (and roots) confined and manageable, while free-growing plants are difficult to control. Moreover, ponds do not usually contain enough rooting medium for gross feeders like water lilies.

PLANTING A BARE-ROOTED PLANT

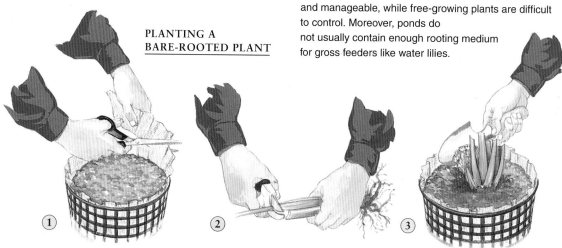

▲ *The following procedure should be adopted for planting out any bare-rooted plants. Buy a generous container; line it with hessian, cut to the correct size (1) and fill it with soil or aquatic compost. Rinse each plant*

thoroughly, keep it moist and trim the old roots. Leave young (usually lighter-coloured) roots in place, even if they are incomplete but seem otherwise healthy, and cut off any fully formed leaves (2). Spread the roots of the

◀ *Plant management is made considerably easier if containers are used and placed within easy reach of the edge of the pond.*

▶ *Containers can now be purchased in a wide range of sizes and designs that cater for every type of pond plant.*

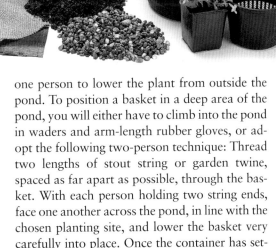

Container-grown plants – both marginal and surface species – come in either pots or baskets. Either way, unless they are specifically sold as "ready-to-install" plants, these pots are likely to be too small. If the container is a plastic pot, transfer the plant to a perforated, hessian-lined basket. This allows water and nutrients to circulate freely around the roots and prevents the plant from rotting in the oxygen-deficient conditions inside a pot. If the plants are to be left in pots for any length of time, a good temporary measure to encourage water circulation is to punch holes around the sides.

Planting techniques for plants removed from their pots or baskets are the same as those for bare-rooted plants above.

When lowering baskets into the pond, do so slowly to avoid disrupting the arrangement or disturbing the fish. If the baskets are intended to rest on the perimeter shelves, these are usually sufficiently shallow and near the pond edge for

one person to lower the plant from outside the pond. To position a basket in a deep area of the pond, you will either have to climb into the pond in waders and arm-length rubber gloves, or adopt the following two-person technique: Thread two lengths of stout string or garden twine, spaced as far apart as possible, through the basket. With each person holding two string ends, face one another across the pond, in line with the chosen planting site, and lower the basket very carefully into place. Once the container has settled in its proper position, gently pull the strings back out through the holes.

(4)

(5)

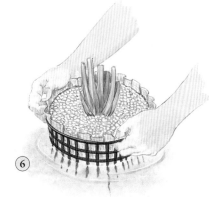

(6)

plant out on the surface of the soil or compost, cover them with added medium (3) and firm this down, ensuring that the crown (growing point) of the plant remains above the surface. Water the plant well (4); this helps get rid of any air in the soil that might dislodge the plant when it is put in the pond. Finally, top off the container with gravel (5) before lowering it into position in the pond (6).

Plant Health and Diseases

ALTHOUGH POND PLANTS ARE VERY RESILIENT, they are still susceptible to attack from various pests and diseases. Unfortunately, because of the danger to fish and other pond life, it is not always possible to spray any affected plants with insecticides or other treatments.

Nevertheless, if sprays can be kept away from the water and a partial water change conducted straight after application, then careful spraying is an option for marginals and bog plants (*never* for surface or floating plants). If the plants are on the pond margin, treat on a calm day, and direct the spray away from the pond. To further reduce the risks, use the minimum recommended dosage. If the plants are in containers, remove them from the immediate pond area before spraying.

Bog and Marginal Plants

The pests that attack these plants are broadly the same as those that afflict normal garden plants, and the methods of controlling them are outlined in standard gardening texts. In addition, observe

▲ *Severe snail damage to a Hosta leaf. Hostas are particularly susceptible to attack by snails and slugs, which make their presence felt from spring onwards.*

● *How can I minimize the risk of introducing pests and diseases carried by plants into my pond?*

Precautionary measures cannot guarantee that all potential health risks are eliminated, but will reduce the threat considerably. Potassium permanganate crystals dissolved at the rate of about 1 tablespoon per 2 gallons (9 litres/2.4 US gallons) of water is an effective disinfectant. Soak plants in this solution for about two hours. Leaves, but not roots, may also be disinfected with a vinegar (acetic acid) solution consisting of about 10ml vinegar per 10 litres of water. This bath should last for only about 5 minutes. Systemic fungicides can also be applied to marginals and bare-rooted plants, allowing enough time for the compounds to be absorbed into the plants' system (a few hours for bare-rooted plants, several days for containerized specimens). All disinfectants should be rinsed off before plants are introduced into the pond. In the case of systemic compounds, soaking bare-rooted plants for several hours in water, or allowing containerized specimens to stand in clean water for about a day, should ensure that they are purged. Finally, check all plants visually and remove any pests or obviously diseased tissues before transferring to the pond.

● *How does thinning out aid plant health?*

When plants are allowed to become pot-bound, their roots become compacted and both their growth and that of the aerial parts becomes restricted. In addition, the ratio of soil to roots and, consequently, nutrients to roots, is reduced. As a result, the plants grow generally weaker and more susceptible to health problems. Moreover, the air between compacted plant stems ceases to circulate freely and can become stagnant and permanently humid. These conditions are ideal breeding grounds for certain plant pathogens, especially fungi.

● *At what time of year are plant pests most prevalent?*

Without a doubt, plant health problems are more common during the growing season. Symptoms may not be particularly evident during the early part of the season, but preventive measures taken at this stage will help control outbreaks later on.

▶ *The China Mark Moth* (Nymphula nymphaeata) *seen here laying its eggs on a* Potamogeton *leaf in late summer, is a major plant pest affecting a wide range of species.*

the special precautions mentioned above. Several other steps may also be taken. For example, if possible, remove all affected leaves and dispose of them well away from the pond. Alternatively, submerge them for a day or two. However, if Irises have been attacked by Leaf Spot (caused by a fungus), these remedies will prove impractical. Submerging the leaves will not work, since the pathogenic agent is not an animal that can be eaten by fish, while removing the leaves could compromise the production of blooms if the attack occurs before the flowering season.

To combat certain pests, a number of safe biological alternatives are available. Such "natural" countermeasures include baiting slugs and snails with fruit or beer.

Surface Plants

The surface plants that suffer most from diseases or pests are water lilies. The following are the most common problems.

Water Lily Aphids. These look identical to the terrestrial blackfly aphid. In the growing season, the problem can be contained by repeatedly playing a jet of water on the affected areas, or submerging the leaves (pads) for a day or so. During autumn, when the females lay their eggs on trees such as the plum and cherry, this pest can be controlled effectively by spraying the bare branches with a proprietary tar–oil compound.

Water Lily Beetles. The larvae of this beetle feed voraciously on lily leaves, eventually destroying the tissues and causing the leaves to disintegrate. Use similar leaf treatment as for aphids and physically remove any beetles and larvae. Cut badly damaged leaves from the plant. As adult beetles hibernate in pondside vegetation, this should be cut back or carefully treated during winter.

Water Lily Crown Rot. This disease causes lily rhizomes to rot into a gelatinous mass. One type affects yellow-coloured cultivars, while another, more recent introduction shows no preference. Early symptoms include blackening of the leaves and progressively smaller leaves. At present, neither type can be cured, and diseased plants must

be discarded. To help avoid re-infestation, drain the pond and – for the first type of rot – treat it with a copper sulphate solution. Against the second type, sodium hypochlorite should be used. It is *vital* that treated ponds be thoroughly rinsed with clean water before fish are reintroduced.

Water Lily Leaf Spot (Brown Spot). There are two types of spot, one that begins around the edges and gradually spreads to the centre of the leaves, and another in which spots appear on the surface of the leaves. Affected leaves should be removed (complete with stem) and disposed of.

China Mark Moth. Two species of the China Mark Moth affect pond surface plants. The larvae of the Brown China Mark feed mainly on the underside of floating leaves, where they also pupate. In the other species, the larvae burrow into the stems, where they hibernate before proceeding to the leaves the following season.

Physical removal of larvae and pupae is a tedious, but necessary, method of control if the infestation is not severe. Alternatively, affected leaves and stems can be removed and discarded.

Other Pests. A wide variety of insects may attack pond and marginal plants at some time. Hosing them off into the pond or submerging the affected leaves are good methods of control, as they exploit the natural tendency of fish to eat any small creatures that fall into the water. While not eradicating the problem, such biological controls may well keep it in check, but only if carried out regularly as part of general pond maintenance.

Bog Plants

BOG PLANTS ARE PROBABLY BEST DESCRIBED as moisture-loving plants; and all bog plants are "marginal" in the sense that they are grown around the edges of ponds and water gardens. However, the distinction between genuinely terrestrial species and bog plants, i.e. between "normal" garden plants on the one hand and, on the other, traditional marginals that have their roots permanently immersed in water, is not always easy to define.

Some types of bog plant, such as the various species and cultivars of *Astilbe* and *Hosta*, only grow successfully in moist ground that is well drained, i.e. not waterlogged. These plants lie at the (relatively) dry end of the "moisture" spectrum. Many of the *Iris* and *Lobelia* species and cultivars, however, also thrive in wetter conditions, including shallow water, and these therefore lie at the other end of the spectrum, where bog conditions merge with the more aquatic environment of the pond itself. The term is usually reserved for those species and cultivars that do best when their roots are submerged for most or all of the year.

Some of the bog plants described here may be displayed for sale undifferentiated from ordinary garden plants. Check before making your final choice.

● *How can containerized bog and marginal plants be protected against being toppled over by strong winds?*

Short plants are generally reasonably stable and will rarely topple over, even in strong winds. Taller plants can be protected in several ways. For example:

1) A heavy stone or brick placed inside the container, i.e. under the rooting medium, will help stabilize it. However, this will reduce the amount of rooting medium accordingly.

2) Heavy pebbles laid on the surface of the medium will also offer considerable, though less, protection.

3) Using broad-based containers will provide greater stability than normal (i.e. plant-pot-shaped) ones.

4) Avoiding over-planting, or preventing tall plants from becoming pot-bound, will reduce the risk of instability.

5) Planting broad and deep (front to back) stands of plants in appropriately large containers will also improve stability.

6) Moving tall plants temporarily into deeper water if stormy conditions are forecast will offer them short-term protection.

7) Erecting a storm break (barrier), such as a slatted fence, also helps. This may be either a permanent structure built in the path of prevailing winds, or a temporary, movable one that can be installed as required. Solid wind breaks should be avoided as they can fall over or be damaged in very windy conditions.

● *How many marginals should I plant around my pond, and of what type?*

The number of marginals that you use really depends on how large an area of moist ground you are able to give over to growing these plants. In certain wildlife water schemes – for example, those where the edge consists of a gentle slope that extends all the way around the pond – you should be able to grow large numbers of exclusively bog plants. On the other hand, where the moist area extends upwards and backwards well into the garden itself, a mixture of bog plants and plants that are capable of growing under somewhat drier conditions may be used almost as traditional (but well-watered) garden plants in "terrestrial" beds bordering the pond. These marginals would include Astilbes and Day Lilies, and even some of the types included later on under the heading 'Shallow-water Marginals', such as Houttuynia. Generally speaking, though, any pond surround will provide some pockets of moist soil and it is the size, number and distribution of these that will normally dictate the concentration of bog plants that can be used. However, you should bear in mind that if you have planted them only in very small numbers, they are not likely to have much effect on the water quality.

▶ *Large garden schemes with extensive moist and shallow-water areas offer excellent opportunities for creating impressive displays at the water's edge.*

Selected Bog Plants

Bugle (*Ajuga reptans*)

This hardy spreading plant is available in several colour forms, ranging from the uniformly coloured, dark (almost greenish/reddish bronze-brown) wild type, to a lighter, tricoloured cultivar. During temperate winters, the leaf rosettes become condensed, as the larger outer leaves are lost, and the stolons (stems with plantlets at their tips) become clearly visible. Uncontrolled, this plant can spread over large areas. It is particularly suitable for growing between the rocks/slabs of an informal pond surround.

Height: Spreading plant with bluish flower spikes, usually around 4in (10cm) tall, but may be taller.

Habitat preference: Full sun. These plants can tolerate both dry and moist conditions, the latter being preferred.

Flowering period: Spring.

Propagation: Splitting of established clumps may be carried out in spring. Individual plantlets or clumps can be planted directly in the soil.

Astilbe (*Astilbe* spp. and varieties)

Astilbes are available in a wide range of shades, mainly from whites/creams, through pinks, to reds. They are all excellent plants for the water's edge, forming large clumps with attractive deeply-divided leaves which are copper-brown during their early stages. While the leaves will fall as autumn approaches, the dried flowering stems remain and can be left on the plants until the following spring. They should, however, be removed before the new-season growth

▲ *Astilbes* (this is Astilbe simplicifolia – Atro rubens) *are versatile plants for the bog garden. The dried-out flowering stems even look good over winter.*

gets under way. Small clumps planted between rocks bordering a pond will spread and join up, resulting in a spectacular display.

Height: From 12–36in (30–90cm) depending on cultivar and ambient conditions.

Habitat preference: Moist (but not waterlogged) ground above the water surface, in full sun or partial shade.

Flowering period: Summer.

Propagation: Division of clumps may be carried out at almost any time of the year, except summer. However, the best time for this is autumn.

Cuckoo Flower (*Cardamine pratensis*)

The Cuckoo Flower is an early-flowering species available in its wild type (single) form and a double one ('Flore Pleno'). Both types are free-flowering, producing lilac blooms above finely cut (almost ferny) leaves. The flower stems – particularly in the wild type – are rather delicate and attractive. They will appear to best advantage when they do not have to compete with gaudily coloured robust neighbours.

Height: Up to 18in (45cm) – slightly less for 'Flore Pleno'.

Habitat preference: Moist ground in full sun or partial shade. Will also grow in very shallow water.

Flowering period: Spring.

Propagation: Wild type may be grown from spring-sown seeds or by division during autumn/winter. The double form does not set seeds and can only be propagated by division of clumps.

Bugbane (*Cimicifuga simplex*)

Bugbane is a useful plant to have in the bog garden, owing to its attractive flower spikes and its tendency to continue flowering late into the outdoor season. It can, however, grow quite tall and must therefore be appropriately located for maximum effect. The common name is a reference to the unusual, slightly unpleasant but not overpowering fragrance of the flowers, which does not necessarily act as an insect repellent!

Height: Up to 4ft (1.2m).

Habitat preference: Moist areas, including those in partial shade.

Flowering period: Summer and into late autumn in some areas.

Propagation: Seeds may be sown in spring; alternatively – and for more immediate results – clumps may be split either in autumn or spring.

Pond Spurge (*Euphorbia palustris*)

This close relative of the garden spurges has, basically, the same greenish-yellow flower heads of many of its cousins. These blooms are produced in great profusion, forming an impressive display during spring

▲ *The Giant Brazilian Rhubarb* (Gunnera manicata) *is a truly impressive plant. There is also a more compact, hardier species,* Gunnera scabra.

and early summer. As the season progresses, the colour of the bright green foliage begins to darken and can become bronze-coloured.
Height: Up to 36in (90cm), but often shorter.
Habitat preference: Moist soil in full sun.
Flowering period: Primarily spring; occasionally also early summer.
Propagation: Clumps may be divided during autumn and winter, but seem to do best when divided in spring as new growth is about to get under way.

Giant Rhubarb (*Gunnera manicata*)
More correctly referred to as the Giant Brazilian Rhubarb, this plant lives up to its name and is only suitable for large displays. The huge flower spikes set seed readily and are as much a feature of the species as its leaves. Although generally hardy, the crowns should be protected during the winter months in areas that experience harsh frosts or snow. The traditional way to do this is by covering the crowns with several layers made up of the previous season's leaves.
Height: Can reach 6.5ft (c.2m) or more under good

growing conditions, but is often smaller than this.
Habitat preference: Moist ground in full sun or partial shade.
Flowering period: Late spring to summer.
Propagation: Seeds can be sown as soon as they are fully ripe. Alternatively, clumps can be split in spring.

Day Lily (*Hemerocallis* varieties)
The Day Lily, while being potentially invasive if left unchecked, is a superb waterside plant owing to the masses of flowers produced on a long-running basis during mid-season. Each flower – as indicated by the common name – lasts for only about one day, but so many are produced that a sizable clump will always have flowers at different stages of maturity. The Day Lily is therefore, in many ways, the water garden's equivalent to the Hibiscus. Many cultivars are available, including the gorgeous double one known as 'Flore Pleno'.
Height: Up to 4ft (1.2m).
Habitat preference: Moist soil or more "normal" garden conditions – as long as these are not overly dry – in full sun.
Flowering period: Summer, extending towards early autumn.
Propagation: Clumps may be divided either in spring or autumn. The latter is preferable.

▲ *Hostas (Hosta spp. and varieties), once sold almost exclusively as normal garden plants, are now widely available for more moist environments.*

Hosta (*Hosta* spp. and varieties)

Hostas are fast-spreading, clump-forming plants that will enhance any waterside arrangement. However, they must be adequately protected from slugs and snails which find these plants particularly delectable. Frequent "baiting" around the plants, preferably using natural methods, such as half grapefruits or small sunken cups (or other vessels) containing beer or ale, should be carried out from spring onwards. Failure to do this can result in badly chewed leaves which can become something of an eyesore, particularly during the summer season. There are numerous types of Hosta, including some strikingly marked variegated hybrids and varieties.

Height: From around 12in (30cm) or less, to over 39in (1m), depending on variety.
Habitat preference: Moist ground in sun (but best avoiding fully exposed sites) or partial shade.
Flowering period: Summer.
Propagation: Division of clumps during autumn or spring. If propagation is delayed until the new growths begin to appear in spring, it is possible to prise these away from the main central area without having to lift the complete plant.

Iris (*Iris* spp. and varieties – but see also under Marginal Plants)

Irises are popular elegant plants that produce colourful displays, particularly from early to mid-summer. The two species featured are regarded as bog plants here because they dislike permanent inundation, preferring instead to be grown either under "normal" garden conditions, or in moist soil

Iris ensata (also referred to as *I. kaempferi*) – the Clematis-flowered Japanese Iris – dislikes limy soils and, particularly, immersion during the colder months of the year. Numerous cultivars are available.

I. sibirca (the Siberian Iris) grows just as well in dry as in moist conditions and tolerates immersion better than *I. ensata*. Numerous cultivars are also available.
Height: Up to 36in (90cm).
Habitat preference: Full sun in dry or moist soil, with periodic inundation tolerated.
Flowering period: Summer.
Propagation: Either by means of spring-sown seeds, or by division of clumps, either in the autumn, or immediately after flowering, i.e. late summer.

Ligularia (*Ligularia* spp.)

Two species of *Ligularia* are especially popular for pond-edge planting: *L. dentata* (also referred to as *L. clivorum*), which has broadly pyramidal inflorescences, and *L. przewalskii* which has tall flower spikes. In *L. dentata*, the leaves are roundish, while in *L. przewalskii* they are much more finely divided or lobed. The flowers of both species are daisy-like, but deep-gold in *L. dentata* (varying in intensity, depending on cultivar) and brighter yellow in *L. przewalskii*. Both species tend to wilt in hot weather.
Height: *L. dentata*, up to 4ft (1.2m); *L. przewalskii*, up to 6.5ft (c.2m).
Habitat preference: Moist ground in full sun or partial shade.
Flowering period: Mid- to late summer.
Propagation: Unlike *L. przewalskii*, *L. dentata* cultivars do not 'breed' true, but may still be grown from spring-sown seed. Both species may be propagated by clump division, either in autumn or early spring.

Lobelia (*Lobelia* spp. and varieties)

Lobelias are among the most striking bog plants, especially the red cultivars of *Lobelia cardinalis*. *L. vedrariensis* (sometimes featured as *Lobelia × gerardii* 'Vedrariensis') and its cultivars and hybrids tend to be more purple and are hardier than *L. cardinalis*, which should receive some protection during particularly cold periods in temperate zones, either by mulching or by transferring some containerized plants to frost-free quarters.

The red-flowered varieties Lobelia 'Cherry Ripe' and 'Will Scarlet' are particularly attractive, as is the violet/purple 'Russian Princess'.
Height: Up to 39in (1m).
Habitat preference: Moist soil during the outdoor season, with drier conditions (preferably frost-free in the case of half-hardy types) during the winter months. Dry conditions do not suit *L. vedrariensis*.
Flowering period: Mid- to late summer.
Propagation: This is usually by means of division in early spring. Allow the plantlets to root well in pots prior to moving to their permanent positions.

▲ *Purple Loosestrife* (Lythrum salicaria) *by the side of a pond. Though this plant is attractive and hardy, it can become invasive if it is neglected.*

▲ *Monkey Flowers* (Mimulus *spp. and cultivars*) *give a splash of bright colour. The more delicate types may require winter protection in some temperate zones.*

Creeping Jenny (*Lysimachia nummularia*)

This spreading plant can form carpets that, as the name implies, creep over minor "obstacles" such as low pondside rocks. As the stems extend over the ground, roots are sent freely from the leaf nodes and these can help bind the soil that can otherwise roll or be washed into the water from informal pond edging. Two forms of Creeping Jenny or Pennywort are widely available, a green-leaved type and a golden one (*L. nummularia* 'Aurea') whose leaves are almost the same colour as the yellow flowers that are produced in abundance by this attractive plant.

Height: About 2in (5cm).

Habitat preference: Moist soil in either sunny or shady positions.

Flowering period: Late spring and into summer.

Propagation: As long as the soil is kept moist, rooted cuttings can be taken at any time of the year but will grow fastest during late spring and early summer. Unrooted stem cuttings also take readily during the growing season if they are not allowed to dry out.

Purple Loosestrife (*Lythrum salicaria*)

This is an impressive plant, especially when a large stand is in full bloom. In the wild type (the form that is best for wildlife set-ups), the tall flower spikes are purple-pink and can attain a height of up to 6.5ft (c.2m), while, in the various cultivars, they tend to be lighter and, often, a little shorter. One of the features that makes *L. salicaria* so attractive as a wildlife waterside plant is that it attracts bees, several types of butterflies and other insects.

Height: 36in–6.5ft (90cm–2m).

Habitat preference: Moist soil; sun or partial shade.

Flowering period: Summer.

Propagation: This is best achieved by division of established clumps during autumn.

Monkey Flower or Musk

(*Mimulus* spp. and varieties)

There are several species and many different *Mimulus* cultivars available, some more winter-tolerant than others in temperate zones. One of the hardiest species is *Mimulus luteus* (Yellow Musk) which is very vigorous and can become invasive, especially since it self-seeds freely. On the more delicate side, *M. cardinalis* (Cardinal Monkey Flower) will require winter protection in cooler zones; alternatively, it can be grown annually from seed. Cultivars vary in hardiness, with some surviving cold temperatures more readily than others, so check for suitability if you wish to over-winter a particular variety *in situ*.

Height: Around 12in (30cm) for *M. luteus* and many cultivars, and up to 24in (60cm) for *M. cardinalis*.

Habitat preference: Moist soil (but not waterlogged or under water for *M. cardinalis*) in full sun.

▲ *The different kinds of Globe Flower* (Trollius *spp.*) *provide a splash of bright yellow in springtime. Shown here is a stand of* Trollius chinensis *in full bloom.*

Flowering period: Summer.
Propagation: Propagation is either by replanting of overwintered leaf rosettes in the case of hardy types, or from spring-sown seed in others. Stem cuttings taken during the growing season will also root readily if kept moist.

Primulas (*Primula* spp. and varieties)

There are many Primulas or Primroses that are suitable for the moist areas around a pond. If they are carefully chosen, they can provide colour from spring into summer. Among the earlier bloomers is the pinkish- or purplish-flowered, spherically-headed *P. denticulata* (Drumstick Primula), which is also available in white (*P. d.* var. Alba). The latest flowering species is *P. florindae* (Himalayan or Giant Cowslip), which is considerably taller, with much looser sulphur-coloured inflorescences. The most unusual species is *P. vialii* (Orchid Primula) which produces conical flower spikes that are darker (redder) at the tips (where the individual blooms are yet to open) and purplish-blue lower down.
Height: About 18in (45cm) for short types; 24in (60cm) for the next; 36in (90cm) for *P. florindae*.
Habitat preference: Moist soil (although *P. florindae*

can tolerate standing water) in sunny positions or partial shade.
Flowering period: *P. denticulata*, *P. rosea* and others flower in spring; *P. alpicola*, *P. aurantiaca*, *P. beesiana*, *P. bulleyana*, *P. japonica*, *P. prolifera*, *P. vialii* and others, in early summer, followed by *P. florindae*.
Propagation: This is best achieved either by early spring division of established clumps, or by late spring/summer-sown seeds collected as soon as they ripen (otherwise, self-seeding is likely to occur). Summer-sown plantlets may require protection in areas that experience severe winter conditions.

Globe Flower (*Trollius* spp. and varieties)

There are several kinds of Globe Flower available for ponds. These range from the tightly-flowered, bright yellow *Trollius europaeus*, through the looser *T. pumilus* with its contrasting colours of bright yellow internal bloom surfaces and dark red or crimson exteriors, to the yellow/orange *T. chinensis* and the more intricately-flowered cultivars such as the various *Trollius × cultorum* hybrids. All are impressive plants in their own distinctive ways.
Height: Around 6in (15cm) for *T. pumilus* and up to 18–24in (46–60cm) for the cultivars and *T. europaeus*.
Habitat preference: Moist ground; sun or light shade.
Flowering period: Spring to early summer.
Propagation: Cultivars should be divided during autumn. Species may be similarly divided or propagated from early autumn-sown seeds.

Shallow-Water Marginal Plants

AT THE SHALLOWEST END OF THE SHALLOW-water range, marginals merge with bog plants – in contrast to the zone preferred by deep-water marginals, aquatics and floating plants. Shallow-water marginals include representatives of genera such as *Iris*, some of which are suited to moist soil (bog) and even drier conditions; these appear under both categories. Others are represented by species and varieties that need to be covered by water for most or all of the year.

At the deeper end, the demarcating line between shallow-water and deep-water marginals is just as difficult to define. Species and varieties which appear under one category in this book may belong to a different one in another. For this book, the demarcating line is 6in (15cm) water depth. If a species or variety can tolerate or thrive in conditions which begin even within the moist soil/shallow-water zone, but can also extend beyond the 6in (15cm) limit, you will find it under Deep-water Marginals, pages 144–147.

● *Can dwarf versions of normally large plants be produced by restricting the size of the container they are grown in?*

Restricting root growth may result in correspondingly restricted overall growth, at least for a time. However, using this technique to house large plant species in small ponds is counterproductive, since the smaller versions produce fewer blooms and are generally inferior to the more freely-growing types. Genuine dwarfing is, in any case, impossible to achieve under normal pond conditions, with the result that plants forced to grow under restricted conditions will either deteriorate to the extent that they need replacing, or break out and overrun the available space. If small plants are required, the best thing to do is to choose either small species (e.g. *Typha minima*) in case of reedmaces, or some of the less vigorous cultivars.

▼ *Carefully chosen marginal plants will produce a long-lived, constantly-changing, spectacular display from spring onwards.*

Selected Shallow-water Marginals

Sweet Flag (*Acorus calamus*)

This is an attractive old favourite that is available either in a green (wild type) or variegated form (*Acorus calamus* 'Variegatus'), the latter being more popular than the former. Although the leaves are Iris-like in appearance, the flowers could not be more different. The most charitable thing that could be said of Sweet or Myrtle Flag inflorescences is that they are far from spectacular. The narrow, pointed leaves give off an interesting citrus-like smell when cut or damaged, and young leaves, particularly in the spring, have a rosy tinge that disappears as they grow.

This is a tough, vigorous plant that dies back almost to ground level during cold winters.

Height: Up to 4ft (1.2m) for the wild type, but usually shorter.

Habitat preference: Full sun or partial shade in conditions, ranging from moist soil, down to around 6in (15cm), or even a little deeper.

Flowering period: Mid- to late summer.

Propagation: Division of clumps, usually recommended during the spring and summer, but also possible during the autumn, once active growth has stopped for the year.

Water Plantain (*Alisma* spp.)

Of the two species that are generally available, the more popular one is *Alisma platago-aquatica* (Water Plantain or Great Water Plantain), which has considerably wider leaves than those of *A. lanceolatum*. These plants are grown primarily for their foliage, rather than their wiry inflorescences of small whitish flowers, which can look a little untidy once flowering is fully under way. Alismas are hardy species that seed freely and can therefore become invasive.

Height: Inflorescences can attain as much as 36in (90 cm) in height.

Habitat preference: Moist soil down to around 6in (15cm), in full sun.

Flowering period: Summer.

Propagation: These species are easy to propagate from seeds as soon as they ripen in late summer, or by division of clumps at any time during the growing season. Autumn division is also possible, but no growth will occur in frost/cold-prone regions until the following spring.

Bog Arum (*Calla palustris*)

Although the scientific name of this delightful plant is similar to that of the Marsh Marigold (*Caltha*

palustris; see Deep-water Marginals, page 145) it couldn't be more different. In the Bog Arum, the "flower" (as in all arums) is not the showy white "petal" that first catches the eye, but the spike in the centre, which, on closer examination, reveals a large number of tiny blooms. The "petal" is, in fact, a spathe or modified enveloping leaf, as found in other well-known plants, such as the Cuckoo-pint or Lords-and-Ladies (*Arum maculatum*) – common in shady, moist, wooded or hedged areas – or the larger and more spectacular Skunk Cabbages (*Lysichiton* – see page 141) and Arum Lily (*Zantedeschia aethiopica*; see Deep-water Marginals, page 147). The spathe in the Bog Arum, however, does not fully envelop the spadix (flower spike), but folds open to expose the tiny yellow blooms.

Height: Up to 12in (30cm).

Habitat preference: Moist soil down to around 2in (5cm) of water, in full sun.

Flowering period: Mid-spring to early summer (a little later if the spring is on the cool side).

Propagation: Can be grown from newly-ripe seeds sown in late summer, or by division of established clumps in spring. Autumn division is also possible, but no new growth will emerge until the spring.

Brass Buttons (*Cotula coronopifolia*)

Also known as Golden Buttons, *Cotula coronopifolia* is a free-seeding plant that usually dies after flowering. Specimens which remain underwater during winter may, however, survive and regrow the following spring if conditions have not been severe. Healthy specimens produce a profusion of bright yellow button-like flowers which, as they wilt, should be removed on a regular basis to avoid uncontrolled seeding and keep the clump looking attractive.

Height: Up to 8in (20cm).

Habitat preference: Moist soil down to around 4in (10cm) of water, in full sun or partial shade.

Flowering period: From late spring to late summer or even early autumn.

Propagation: This plant grows easily from spring-sown seeds. If sowing is carried out in early spring in frost-prone areas, some protection should be provided for young plants.

Umbrella Plant (*Cyperus alternifolius*)

Several species of *Cyperus* are frequently used as marginals, three of which are widely available: *C. longus* (Sweet Galingale), a good plant for temperate wildlife schemes; *C. papyrus*, the well-known Papyrus (see Deep-water Marginals, page 145); and *C. alternifolius* (Umbrella Plant), often grown as a house plant. Provided that the shoot tips are well covered with water, and provided conditions are not excessively harsh for a prolonged period of time, *C. alternifolius* is an adaptable plant that is winter-hardy, even in temperate zones.

Height: Up to around 4ft (1.2m).

Habitat preference: Moist soil down to c.6in (15cm) or even slightly deeper, in full sun or partial shade.

Flowering period: Summer to early autumn.

Propagation: Mainly by division of clumps during spring.

Cotton Grass (*Eriophorum angustifolium*)

When a large stand of Cotton Grass is in full bloom, it makes a truly impressive spectacle. However, large stands are only feasible if plants are grown in open ground or beds, rather than in containers. One potential problem with doing this is that Cotton Grass is highly invasive and can prove difficult to control.

◀ *The cottony perianth (sepals and petals) of Cotton Grass* (Eriophorum angustifolium) *becomes most conspicuous after flowering.*

◀ *(Far left) The creamy yellow spadices (flower spikes) of the Bog Arum* (Calla palustris), *combined with the white spathes, make it a highly desirable plant.*

▲ *Water Irises* (Iris laevigata) *look their best when they are planted in large clumps. Not all varieties will grow in deep water – be sure to check.*

▲ *The Corkscrew Rush is an unusual form of* Juncus effusus *that requires occasional monitoring to remove any straight leaves that appear.*

This species requires an acid rooting medium, such as peat, which can be easily provided in measured doses if the plant is containerized.

Height: Up to 18in (46cm).

Habitat preference: Peaty/acid conditions in moist soil or water down to around 2in (5cm) in depth, in full sun.

Flowering period: Summer.

Propagation: Late summer sowings of newly-ripened seeds will germinate, but unpredictably. The easiest method of propagating this hardy, fast-spreading plant is by clump division, either in autumn or spring.

Houttuynia (*Houttuynia cordata* and varieties)
Houttuynia cordata is available in three main forms: the wild type which has deep green, heart-shaped leaves and simple white flowers, a semi-double variety known as *H. cordata* 'Plena' and *H. cordata* 'Chameleon' or 'Variegata', which has colourful leaves that contain green, creamy yellow and red and is the most popular type.

Houttuynia is an invasive plant that gives off a distinctive smell when handled and, particularly, when stems or roots are cut during routine maintenance or propagation.

Height: Usually from 6–18in (15–45cm), but can grow a little taller.

Habitat preference: Moist soil down to around 2in (5cm) of water, in full sun or partial shade.

Flowering period: Late spring, through summer.

Propagation: Established clumps may be divided either in autumn or spring, with the former not showing any new growth until the new season.

Water Irises (*Iris laevigata, I. versicolor* and varieties)
Iris laevigata (Japanese Water Iris) is one of two species that are commonly available for the shallower areas of the pond. The similar *I. versicolor* (American Water Iris) has slightly narrower petals. Several cultivars of each species are also widely available, ranging from types with white, through mottled, to deep purple-blue flowers and – in the case of *I. laevigata* 'Variegata' – with variegated leaves.

Although the normal number of petals in a flower is three, some cultivated varieties have double this number. When buying Irises, you should check whether the variety is a hybrid with *I. ensata* (see Bog Plants, page 134), since such plants are less tolerant of total submersion.

Height: *I. laevigata*: up to 40in (c.1m) for some cultivars, but shorter for 'Variegata'; *I. versicolor*: around 30in (75cm).

Habitat preference: Moist soil down to around 4in (10cm), or a little deeper for *I. versicolor* – but see above for 'ensata' hybrids. A situation in full sun is preferred by all types of Water Iris.

Flowering period: Early to mid-summer for most types; *I. versicolor* may begin flowering a little later than *I. laevigata*.

Propagation: The most effective method is by division of established clumps once flowering has been completed. Spring-sown seeds of the two wild types will germinate readily. Cultivars will not, however, grow true to variety.

Corkscrew Rush (*Juncus effusus* 'Spiralis')

Most *Juncus* species are straight-leaved plants which, while available for ornamental ponds, are probably best for wildlife schemes. The Corkscrew Rush is an exception. As its name implies, the characteristic needle-like rush leaves are contorted, rather than straight, giving the plant an unusual and more spread-out habit than its wild counterparts. Most specimens still harbour a latent genetically-based tendency to produce occasional straight leaves. These should be removed as soon as detected if the corkscrew leaf arrangement is to be maintained.

Height: Up to 18in (45cm).

▲ *The broad leaves of Skunk Cabbage (this example is* Lysichiton camtschatcensis, *Asian Skunk Cabbage) make them strong subjects for the water's edge.*

Habitat preference: Moist soil down to around 2in (5cm) of water in sunny locations.

Flowering period: Summer.

Propagation: Established clumps may be divided, either in autumn or spring. If division is delayed until spring, just as new growth is beginning to appear, it is easier to spot and discard straight-leaved offsets.

Skunk Cabbages (*Lysichiton* spp.)

There are two species of Skunk Cabbage available, the yellow-flowered *Lysichiton americanus* (American Skunk Cabbage) and the white-flowered Asian version, *L. camtschatcensis*. A hybrid is also occasionally on offer. Despite their common name, Skunk Cabbages are not foul-smelling at all, but are, rather, beautiful, desirable plants that produce an impressive display in spring. Plants may prove a little slow in flowering, especially if bought as young specimens, but they are well worth persevering with.

An alternative version of the generic name, *Lysichitum*, is also frequently encountered in water-garden literature.

Height: Around 4ft (1.2m) for *L. americanum* and a little less for *L. camtschatcensis*.

Habitat preference: Moist soil down to around 2in (5cm) of water in full sun or partial shade.

Flowering period: Early to late spring.

Propagation: This is usually achieved by means of newly-ripened seeds which can take until mid-to-late summer to mature.

Water Mint (*Mentha aquatica*)

Water Mint was once one of the most popular marginal plants both for ornamental and wildlife schemes. It is still as widely available as ever, but its application has been gradually shifting towards wildlife arrangements, probably as a result of the growth of this sector of the hobby and the wider availability of new and more colourful plant species and cultivars.

Water Mint is an invasive plant with deliciously aromatic leaves. The pinkish/lilac inflorescences are variable in size and abundance and are attractive to insects. At the height of the season – but prior to flowering – vegetative growth is particularly rapid and may need regular monitoring and controlling. Once the flower shoots begin to appear, this growth slows down markedly.

Height: Up to 18in (45cm) or a little taller.

Habitat preference: Moist soil, down to around 6in (15cm) of water, in full sun.

Flowering period: Summer.

Propagation: This is very easy, as even the smallest lengths of stem will root, almost at any time during the growing season, as long as they have at least one leaf node. Clumps may also be divided either in spring or autumn.

Bog Bean (*Menyanthes trifoliata*)
Also known as the Buckbean or Marsh Trefoil, this old favourite is used widely, both in ornamental (largely informal) and wildlife ponds. The three-lobed leaves (hence the designation "Trefoil") stand clear, but close to the water surface, facing upwards on shortish, elegantly curved stems, while the flower heads are taller and carry delicately fringed whitish/pinkish/purplish flowers. This is a spreading plant that looks particularly good along the shallow-water edges of more informal schemes.
Height: Around 10in (25cm).
Habitat preference: Moist soil down to around 6in (15cm) of water, in full sun or partial shade.
Flowering period: Spring to early summer, but may continue into mid-summer.
Propagation: Individual stems will root in spring, as long as they contain at least one bud. Alternatively, clumps may be divided in the usual way in spring.

Water Forget-Me-Not (*Myosotis scorpioides*)
For best effect, the Water Forget-Me-Not should be planted in sizable clumps. Otherwise, the delicate flowering stems can become a little lost if surrounded by more robust, larger-bloomed plants. This species is sometimes reported as being unreliably perennial. However, this applies to areas that experience particularly severe winters. Elsewhere, it is as reliable as most other perennials – and, indeed, more reliable than many. The Water Forget-Me-Not is an attractive "invader" that needs to be kept under control to

prevent it from overrunning other slower-growing shallow-water and bog plants. Several varieties are available, including the free-flowering *M. scorpioides* 'Semperflorens'.

The Water Forget-Me-Not still frequently appears under its old name, *Myosotis palustris*, in pond and water-garden literature.
Height: Up to 8in (20cm).
Habitat preference: Moist soil down to around 6in (15cm) in full sun or partial shade.
Flowering period: Summer.
Propagation: Spring-sown seeds will germinate if protected from severe weather. Clumps may also be divided in spring, preferably retaining only the younger stems, which will sprout readily and grow into substantial plants in a single season.

Greater Spearwort (*Ranunculus lingua*)
Also known as the Giant Spearwort, *Ranunculus lingua* is an invasive waterside plant that produces bright yellow buttercup-like flowers. The wild type is particularly popular among wildlife enthusiasts, while a cultivated form, *R. lingua* 'Grandifloris' (alternatively available as *R. lingua grandiflora*), is more ornamental and possesses larger flowers. The leaves on non-flowering shoots are heart-shaped, while those on flowering stems are narrower and more pointed, reflecting the "Spearwort" part of the species' common name.
Height: Up to 36in (90cm).
Habitat preference: Moist soil down to around 6in (15cm) of water, in full sun or partial shade.
Flowering period: Late spring and into summer.
Propagation: Although propagation from newly-ripened seed is possible, the usual method is by division of established clumps in early spring.

Arrowhead (*Sagittaria* spp. and varieties)
There are several species and varieties of Arrowhead widely available, the most commonly encountered being: *Sagittaria japonica* (Japanese Arrowhead), *S. latifolia* (American Arrowhead, Dutch Potato or Wapato) and *S. sagittifolia* (Common Arrowhead). The double forms, which are particularly impressive, carry 'Plena' or 'Flore Pleno' after the species name.

All require basically the same conditions and all produce swollen stem tips (overwintering buds or turions) as autumn approaches. In the wild, these eventually float away in autumn/winter and become established as new plants at the beginning of the following season.
Height: Between 18–24in (45–60cm) for most types, except *S. latifolia*, which can grow up to 5ft (1.5m).
Habitat preference: Down to around 6in (15cm) or a little deeper, in full sun.
Flowering period: Summer.

▲ *The delicately fringed blooms of the Bog Bean* (Menyanthes trifoliata) *make a lovely informal display at the edge of a shallow pond.*

Propagation: During the growing season, clumps may be divided and replanted. Once the season is over, the swollen turions may be planted in new locations and will establish themselves quickly in spring.

Zebra Rush
(*Schoenoplectus lacustris tabernaemontani*)
There are two forms of this plant, the Striped Rush (*S. l.t.* 'Albescens'), formerly known as *Scirpus albescens*, and the Zebra Rush (*S. l.t.* 'Zebrinus'), formerly known as *Scirpus zebrinus* or *Scirpus tabernaemontani (tabernaemontae) zebrinus*.

Both types are white/cream and green, but in different arrangements. In 'Albescens', the stems are

predominantly whitish green, with longitudinal green streaks. In 'Zebrinus', the stems are banded across in white/cream and green. Both plants are spectacular but invasive and must be kept in check. 'Zebrinus' is also susceptible to "reverting to type" – producing green leaves. These must be removed to maintain the banded appearance.
Height: 'Albescens', up to 4ft (1.2m); 'Zebrinus', up to 36in (90cm).
Habitat preference: Moist soil down to around 6in (15cm) of water, in full sun or partial shade.
Flowering period: Summer.
Propagation: Established clumps may be divided either in spring or during the growing season.

Dwarf Reedmace (*Typha minima*)
There are three *Typha* species and one cultivar for ponds (see Deep-water Marginals, page 146). Only one – the Dwarf Reedmace/Bulrush – is a shallow-water marginal suitable for small schemes, though it also looks great around large ponds. *Typha minima* is not as invasive as the larger types, but will still spread, and so is best containerized. The main attraction of all reedmaces is the characteristic fruiting bodies produced after flowering. These are the female parts of the spike-like inflorescence (spadix), which appear below the male parts that form the elongated tip. Together they form the typical bulrush "head".
Height: Around 18in (45cm), but can be taller.
Habitat preference: Moist soil down to around 6in (15cm) in full sun.
Flowering period: Summer.
Propagation: Established plants may be divided either during the autumn or spring.

Brooklime (*Veronica beccabunga*)
A large stand of Brooklime, with its abundance of tiny flowers, is a colourful contribution to any pond, particularly to informal and wildlife schemes. This freely-seeding, -rooting and -spreading species quickly covers large areas. If control is necessary, it is best to containerize this plant and remove any seedlings that may appear in the moist areas around the pond.
Height: Up to 8in (20cm), but often less than this.
Habitat preference: Moist soil down to around 4in (10cm) of water, in full sun or partial shade.
Flowering period: Summer.
Propagation: Established plants may be split either in the autumn or spring. Stem cuttings will also root freely from late spring onwards and seeds will germinate at any time during the growing season.

◄ *The Dwarf Reedmace or Bulrush* (Typha minima) *is suitable for all types of arrangement, but particularly so for small- to medium-sized schemes.*

Deep-Water Marginals

DEEP-WATER MARGINAL PLANTS ARE DEFINED here as those that will grow and thrive in water more than 6in (15cm) deep. Among this group are a number of species and varieties that will grow equally well not just at shallower depths, but also in the moist conditions enjoyed by bog plants. Some of these, for example the Arum Lily (*Zantedeschia aethiopica*), can even be grown as pot or garden plants. However, because these plants are tolerant of deep-water conditions, they have been included in this section.

Q&A

● *When is the best time to split marginal plants?*

... Most species and varieties can be divided during their dormant period. However, the majority can also be split during the growing season, since they do not face the problem of dehydration that affects normal garden plants. Plants split during the growing season will often not flower as profusely during the first few months as they would do if left undivided, but should return to normal the following season. Spring division – when growth is beginning to get under way – is particularly beneficial for some species and varieties, e.g. *Pontederia cordata* (Pickerel Weed), while post-flowering splitting is recommended for others, e.g. Iris species and varieties. Specific details are included for each plant featured in the bog and marginal selections.

● *Why can't double flowers and cultivars be propagated from seed?*

In the case of double forms, the reason is quite simple: they do not set viable seeds. In the case of cultivars and hybrids, many can actually be propagated from seed. However, the resulting "crop" does not always grow true to form, while others produce plants that are not as robust as their parents.

● *Should marginals be overwintered indoors?*

Only the more tender types require or will benefit from this treatment, and then only in temperate zones with severe winter conditions. Hardy types are best left to spend a period of winter dormancy outdoors, irrespective of the weather.

Selected Deep-water Marginals

Flowering Rush (*Butomus umbellatus*)
Despite its common name, *Butomus umbellatus* is not a rush with insignificant inflorescences, but a member of the small family Butomaceae, which contains only five genera. The flowers are pink and are held well above the leaves. Although the family contains a few tropical representatives, *B. umbellatus* dislikes hot climates. A white-flowered form, referred to as *B. umbellatus alba* or *B. umbellatus* 'Alba' is also occasionally available.
Height: Up to 4ft (1.2m).
Habitat preference: This versatile plant grows in moist soil or water depths down to more than 12in (30cm), in full sun or partial shade.
Flowering period: Summer.
Propagation: Divide established clumps at any time during the growing season, although autumn division is also possible. In addition, bulbils (small bulb-like structures) that appear at plant/soil level may be removed and replanted in spring.

Marsh Marigold (*Caltha palustris*)
Various types of the Marsh Marigold or Kingcup
are available. The most basic (original) type is the
yellow-flowered, single *C. palustris,* one of the earliest waterside bloomers in temperate zones. There is
also a white variety, *C. palustris* var. *alba*, which has
white petals with a golden centre, and a double gold-
en form known either as *C. palustris plena* or as
C. palustris 'Flore Pleno'. Occasionally, a fourth,
larger, Kingcup – the Himalayan Marsh Marigold –
is also seen; it is referred to as *C. palustris* var.
palustris.
Height: *C. palustris*: up to 24in (60cm); 'Flore Pleno'
and *alba*: up to 12in (30cm); *C. palustris* var. *palustris;* up to around 40in (c.1m), but usually shorter.
Habitat preference: *C. palustris* and *C. palustris* var.
palustris can grow in anything from moist soil down
to 12in (30cm) of water in full sun. *C. palustris* var.
alba prefers shallower water, while 'Flore Pleno'
prefers conditions ranging from moist soil down to
around 2in (5cm). It is thus a shallow-water marginal, but is included here simply because it is a member
of the same species as the others.
Flowering period: Early to late spring, often with a
second, less vigorous, flowering during summer.
Propagation: Sow seeds as soon as they are ripe
('Flore Pleno' does not set seed), usually during the
earlier part of summer. Mature clumps may be divided, either during the growing season, or in autumn.

Papyrus (*Cyperus papyrus*)
This is the legendary Papyrus of ancient Egypt and a
close relative of the Umbrella Plant (*Cyperus alternifolius*; see Shallow-water Marginals, page 139).
Papyrus is not fully frost-hardy. Therefore, in ice-
prone temperate zones, it should be grown as deep as
possible. Owing to its great height, Papyrus is also
susceptible to damage from strong winds.
Height: Up to 16ft (5m), but usually considerably
shorter than this.
Habitat preference: Moist soil, down to around 12in
(30cm), in full sun.
Flowering period: Summer.
Propagation: Division of established clumps during
autumn. Alternatively, seeds may be sown once they
ripen, but young plants must be protected against
cold conditions.

▶ *Marsh Marigolds* (Caltha palustris) *are often seen
growing wild by the side of a stream. All varieties make
good additions to informal water garden schemes.*

◀ *Flowering Rush* (Butomus umbellatus), *which is not a
rush at all, has an attractive inflorescence. Although it
will grow in full sun, it is not suited to a hot climate.*

Manna Grass (*Glyceria aquatica variegata*)
Manna Grass, also known as Sweet Manna Grass or
Variegated Water Grass, appears under at least four
names in pond plant catalogues: *Glyceria aquatica
variegata, Glyceria maxima* var. *variegata, Glyceria
maxima* 'Variegata' and *Glyceria spectabilis variegatus.* This is a beautiful, vigorous, invasive plant that
needs to be controlled by growing in a container,
especially in smaller water schemes. New shoots have
an attractive pink tinge. Manna Grass is a versatile
plant which, in its wild (unvariegated) form, is sometimes used to stabilize stream and river banks.
Height: Up to 4ft (1.2m), but often shorter.
Habitat preference: Moist soil, down to 8–12in
(20–30cm) of water, in full sun.
Flowering period: Summer.
Propagation: The easiest method is by division of
established clumps, either in autumn or early spring.

Yellow Flag (*Iris pseudacorus* and varieties)
The Yellow Flag or Flag Iris is the tallest available
species in the genus for cultivation in ponds. This
wild type has, over the years, been developed into a
number of forms, such as the particularly beautiful
'Variegata', whose variegations begin to fade as the
season progresses. Other cultivars include a double
form, 'Flore Pleno', and *I.p.* var. *bastardii* (also
known as *I.p.* 'Bastardii' or 'Sulphur Queen'), whose

145

▲ *The Pickerel Weed* (Pontederia cordata) *is a versatile plant that can be grown in a wide range of water depths. Several colour varieties are now available.*

blooms are a creamy yellow. Perhaps the most unusual cultivar is *I.p.* 'Holden Clough', which has brown/purple-streaked yellow flowers. The cultivars are all shorter and less vigorous than the wild type.
Height: Up to 4ft (1.2m) for the wild type and between 24–36in (60–90cm) for the cultivars.
Habitat preference: Moist soil, down to 12in (30cm) for the wild type, but shallower for the cultivars, especially 'Holden Clough', which should be treated as a shallow-water marginal. All types like full sun.
Flowering period: Early to mid-summer.
Propagation: The cultivars are all best propagated by division of clumps once flowering is completed. The wild type may, in addition, be propagated via spring-sown seeds.

Water Purslane (*Ludwigia palustris*)

Also known as the Marsh Ludwigia, *Ludwigia palustris* is perhaps more familiar as an aquarium plant, since it grows exceptionally well underwater. However, in recent years, it has begun to make an appearance as a pond plant, particularly in northern European countries. When grown in deep water, the aerial shoots appear to be shorter than when the plant is cultivated in shallower conditions. The flowers, while pretty, are very small, but the leaves are an attractive pinkish/purplish-bronze, particularly in strong sun.

Height: Up to around 20in (50cm), but often shorter.
Habitat preference: Water down to around 12in (30cm), in full sun.
Flowering period: Summer.
Propagation: This is either from spring-sown seeds, from division of clumps during the growing season, or by separating individual underwater shoots (or bunches) containing adventitious roots growing out of the leaf nodes.

Pickerel Weed (*Pontederia cordata*)

This is a vigorous plant that bears narrowly heart-shaped leathery leaves. Pickerel Weed does well in a range of conditions. While being quite hardy when planted in deep water, plants that are cultivated in moist soil or very shallow water should be given some winter protection in ice-prone regions to prevent their roots from freezing.

In addition to the widely available wild type, a few colour varieties have also begun to appear more regularly in recent years, including white, pink and deep mauve forms. A second, taller, species, *P. lanceolata*, which has narrower leaves than *P. cordata*, is also occasionally available.
Height: Around 30in (75cm).
Habitat preference: Moist soil, down to around 8–10in (20–25cm) in full sun.
Flowering period: Mid- to late summer.
Propagation: Divide clumps once growth has begun in spring. Propagation from seeds is also possible, but (unusually) these must be sown while still fresh (green) during late summer.

Reedmace (*Typha* spp.)

Besides the Dwarf Reedmace (*Typha minima*), there are several other members of the genus that can be accommodated in medium-sized and larger ponds. The most modestly sized of these is *Typha stenophylla* (also referred to as *T. laxmanii*), followed by *T. angustifolia* (Narrow-leaved Reedmace) and, finally, by *T. latifolia* (Giant or Great Reedmace). Giant Reedmace is also available in a variegated form, *T. latifolia* 'Variegata', which is shorter and somewhat less robust than *T. latifolia*.

Other names for the Reedmaces are "Cat's-tails" or "Bulrushes", the latter of which should be reserved for the species of *Schoenoplectus* (*Scirpus*) described on page 143.
Height: Up to 6.5ft (2m) for *T. latifolia* (shorter for 'Variegata'; around 6ft (1.8m) for *T. angustifolia* and about 4ft (1.2m) for *T. stenophylla*.
Habitat preference: Moist soil, down to around 12in (30cm), in full sun.
Flowering period: Summer.
Propagation: Division of established plants early in the spring.

Arum Lily (*Zantedeschia aethiopica*)
The Arum Lily is traditionally grown as a normal garden plant in frost-free areas. Some varieties are frost-hardy in temperate zones that do not experience severe winters. In ponds, three varieties are most commonly encountered: the "original" White Arum Lily, the shorter and reportedly tougher, but also white, *Z. aethiopica* 'Crowborough' and the unusual 'Green Goddess', whose 'petals' (actually a spathe, which is a modified leaf) are mainly green around the edges and creamy white towards the centre, from which the yellow flower spike (spadix) emerges. In addition, the tenderer *Z. elliotana*, with gorgeous yellow spathes and cream-speckled dark green leaves, is occasionally available. All are striking.
Height: Up to 4ft (1.2m) for the original type; c.30in (75cm) for the two varieties.
Habitat preference: Moist soil, down to around 12in (30cm) in full sun. In frost-prone areas, planting in deeper water offers greater cold protection.
Flowering period: Spring to summer.
Propagation: Separate young tubers during the resting (late autumn/winter) season, or divide actively shooting plants early in the growing season.

Other Marginals

Other bog and marginal plants not treated in the foregoing pages include the various *Canna* species and varieties, which can be planted in water down to around 8in (20cm); the increasingly widely available Giant Canna (*Thalia dealbata*), which can, reportedly, grow in water as deep as 24in (60cm); and the impressive, but relatively tender, *Colocasia esculenta* varieties, which need water up to 12in (30cm).

There are also a number of plants from tropical or subtropical zones that make surprisingly excellent, marginals. This group includes the various Amazon Swordplants (*Echinodorus* spp.), Bacopas (*Bacopa* spp.) and many others. Such plants are usually sold as submerged plants for tropical aquaria, but in fact they are tropical marginal plants that have both aerial and submerged foliage. You should be able to look up details of these plants quite easily in tropical aquarium texts or specialist plant publications (see Further Reading, page 202).

▼ *The spectacular fluted white Arum Lily* (Zantedeschia aethiopica) *will be equally at home in deep water or in moist terrestrial conditions.*

Surface Plants

SURFACE PLANTS ARE DISTINCT FROM FLOATING plants in having their roots firmly anchored in the substratum. Unlike submerged species, surface plants do not oxygenate the water, but they do provide shade and shelter and absorb chemical compounds via their roots.

There are relatively few surface plant species. The genus *Nymphaea* (water lily) outstrips all other types of pond plants, both in popularity and in the number of varieties available. Other lily genera include *Nuphar* (with only a few representatives) and *Nelumbo*, the lotuses.

With the possible exception of *Nymphoides peltata*, the Water Fringe or Fringed Water Lily, surface plants are grown in containers to prevent them from spreading and to make maintenance as straightforward as possible.

Q & A...

● *Since surface plants are gross feeders, should they be fertilized in any way?*

Just as long as you use good-quality soil or aquatic compost for planting, there should be no need for you to supply the plants with any additional food, at least during the first growing season. The same applies if plants are divided or replanted after a single season. If you still prefer to feed the plants, however, there are a number of excellent slow-release fertilizer tablets on the market that will help both newly divided plants (provided this operation is carried out during the spring or growing season), as well as those that are left in their containers for more than one season. This same general rule also applies to marginal plants.

● *What is the recommended stocking level for surface plants?*

Surface plants, together with floating plants, can cover as much as 60% of the pond surface. If the pond contains an adequate stock of submerged plants, then around 30% cover will be sufficient to help control algae, but this will not necessarily offer adequate protection against some predators or water loss through evaporation.

Selected Surface Plants
Water Hawthorn
(*Aponogeton distachyos* or *distachyus*)
The Water Hawthorn is sometimes known as the Cape Pondweed, in reference to its South African origins. This attractive deep-water surface plant has an edible tuberous rootstock, strap-like surface leaves and distichous flowers (blooms arranged in two diametrically opposite rows, hence the species name). The flowers' vanilla-like fragrance is most pronounced on warm, still evenings. When planted suitably deep, the Water Hawthorn is hardy, even during harsh winters in temperate climes. It is not, however, at its best during long, extremely hot, spells.
A yellow-flowered, tenderer species, *A. desertorum* (*A. kraussianum*), is occasionally available.
Habitat preference: Water to a maximum depth of 36in (90cm), in full sun, is comfortably tolerated by large specimens.

▲ *The hardiness of lotuses (*Nelumbo *species) varies between species and varieties, but none are hardy enough to endure winter in a temperate climate.*

Flowering period: From late spring, well into autumn.
Propagation: This is a freely seeding plant that germinates easily from fresh green seeds (which are collected mainly during early summer). Alternatively, established plants may be divided during early spring.

Lotus (*Nelumbo* spp. and varieties)

The "original" Sacred or Asian Lotus (*Nelumbo nucifera*) is a frost-tender plant that only does well during protracted spells of warm weather. It is therefore only suitable for outdoor cultivation in temperate zones during the hottest months of the year. At other times, it must be protected under glass if it is to survive. *Nelumbo lutea* – also known as *N. pentapetala* – (American Lotus, Water Chinquapin/ Chinkapin, Pondnuts or Water Acorn) is somewhat hardier, but cannot be regarded as being frost-hardy, although it may survive in deep water if winter conditions are not too harsh.

Both species are available in their wild-type forms and in a range of colour varieties, with *N. nucifera* also having been developed into several double cultivars. There are also many commercial hybrids that have been produced between the two species and/or some of the cultivars.

While lotuses are regarded as surface plants, owing to their close relationship with the *Nymphaea* and *Nuphar* water lilies, they behave in a very similar way to marginal plants once established, in that many of their leaves and all of the blooms are held well clear of the water.

Certain varieties, such as 'Momo Botan', are also able to grow in shallow water, at around a depth of 6in (15cm) – just within the domain of the shallow-water marginals. However, in their native habitats, lotuses spread in large beds that can extend well away from the margins towards the centre of large ponds and lakes.

In addition to their impressive blooms (which can look somewhat less than impressive once they begin to go over the top), lotuses are also famous for their large 'pepperpot' seed capsules. The lotus 'roots' – the banana-like tubers – are edible, as are the leaf stems and seeds.

Height: Up to 5ft (1.5m) for *N. lutea* and 7ft (2.1m) for *N. nucifera*. Cultivars and hybrids can be just as tall, or shorter, depending on variety.
Habitat preference: Shorter cultivars such as 'Momo Botan' can grow in water between 6–12in (15–30cm) deep. Others require deeper conditions down to around 30in (75cm). Both species and cultivars require full sun.
Flowering period: Summer.
Propagation: Both wild types may be propagated from spring-sown seeds. Cultivars (plus the two wild types) can be divided in spring.

Pond Lily (*Nuphar* spp. and varieties)

There are several species and numerous varieties of *Nuphar* lilies (also known as Spatterdocks) which are cultivated in ponds. The most famous in northern European countries is *N. lutea* (Yellow Pond Lily or Brandy Bottle – the latter name alluding to the alcoholic smell and bottle shape of the flowers). Less hardy (though still tough), and perhaps better known in its native lands, is *N. japonica* (Japanese Pond Lily), while in the US, *N. avena* (American Spatterdock or Mooseroot) is well known.

The majority of species and cultivars are too large and vigorous for small ponds, but *N. minima* – probably a synonym of *N. pumila* (Dwarf Pond Lily) – will grow in shallower water and is suitable for small and medium-sized schemes.

Numerous varieties, plus a few hybrids of the various species, are available, most bearing yellow flowers, although a few, e.g. *N. japonica* var. *rubrotincta*, have deeper-coloured blooms. Some species, particularly *N. sagittifolia* (Cape Fear Spatterdock), are also sold as aquarium plants.

Habitat preference: Most species and varieties can be cultivated in water around 5ft (1.5m) deep – although *N. lutea* can grow at depths of about 13ft (4m). Smaller types are best cultivated at depths of

▲ *Despite their large size, Pond Lilies (*Nuphar *species and varieties) produce relatively small flowers. Most except the dwarf species require a good-size pond.*

18in (45cm) or less. All cope equally well with full sun or partial shade, and can grow in flowing water, something that *Nymphaea* lilies cannot tolerate.
Flowering period: Summer.
Propagation: Divide established plants in early spring.

Water Lily (*Nymphaea* spp. and varieties)
The water lily (genus *Nymphaea*) is perhaps the best known of all aquatic plants, and is described extensively in specialist literature (see Further Reading, page 202). The following summary provides only the most general overview.

Traditionally, water lilies are divided into two broad categories, hardy types and tropical types. These are further divided into species and hybrids, with the tropical category further subdivided into day bloomers and night bloomers.

Among the hardy species, parentage is often difficult, if not impossible, to trace, and names may therefore appear in slightly different forms in pond and water-garden literature. The varieties themselves are, however, well known and fixed; thus, an 'Attraction' is definitely such, irrespective of any doubts about its lineage.
Hardy Lilies. This group contains plants for every type of water scheme. They range from miniatures like the white, scented *N. pygmaea* 'Alba', the smallest of all water lilies, and the canary-yellow *N. pygmaea* 'Helvola' – both of which can be cultivated even in small water features – to substantial varieties like *N.* 'Gladstoniana', a superb white-flowered lily that is only suitable for large ponds and lakes.

Despite being hardy, dwarf varieties grown in very shallow water should be provided with sufficient depth to ensure that the rhizomes remain unfrozen during the coldest periods of the year. While resistant to the cold, hardy lilies dislike flowing water and splashes, both of which have a markedly adverse effect on growth. These varieties must therefore be kept in still water that is situated well away from fountains, cascades, streams or other water-disturbing pond features.

The leaves (pads) of hardy lilies lie flat on the water surface, except when grown in crowded conditions. The blooms are also located on the surface in all but a few exceptional varieties, such as *Nymphaea* 'Firecrest', which holds its blooms just above water.

Hardy lily flowers, some of which are delicately scented, last for about 3–5 days, opening in the morning and closing as evening approaches. Once a

▶ *The water lily* Nymphaea 'Candidissima rosea' *produces impressive numbers of large pink blooms during the summer season. It can be grown in water down to a depth of around 36in (90 cm).*

flower dies, it sinks. It is a good idea to "dead-head" free-flowering varieties on a regular basis, especially in smaller ponds. This helps reduce the total organic load on the pond filter, which, if left unchecked, can result in a deterioration in water quality.

Tropical Lilies. Tropical lilies are considerably tenderer than their hardy cousins. Indeed, other than in tropical regions, these plants should be seen as strictly spring/summer pond occupants that need to be brought under cover during the coldest times of year.

If the plants are bought as tubers rather than as rooted, sprouting plants, these should be soaked until they sink and then planted in pots or baskets which are subsequently submerged in suitably larger watertight containers in a sunny position indoors. After a few weeks, there should be enough foliage (and the weather should be sufficiently mild) for the baskets to be transferred to their permanent position. If pots

have been used, the sprouting plants should be transferred to baskets before being placed in the pond.

Sprouting tropical lily plants should not be placed in very deep water straight away (an exception to the general advice given earlier), but adapted gradually to their eventual depth. It is also helpful to transfer growing plants to progressively larger containers until full size is attained. This may be labour-intensive, but it appears to help the plants.

In autumn, the baskets or crates should be lifted from the pond and allowed to dry slowly. Once this process has been done, the tubers will have ripened sufficiently and can be lifted out and stored in damp sand over winter under frost-free conditions.

In day-flowering tropical lilies, the blooms open during the morning and close as evening approaches. In the night-flowering types, the blooms open during the evening and close again at sunrise, but may remain open for at least a few hours on cloudy days. There are beautifully scented representatives among both the day and night bloomers. Both may also vary considerably in size and can therefore be matched to suit the available space, though there are fewer small species and hybrids than among the hardy lilies.

The smallest tropical lily is probably the day-blooming white *N. heudelotii* var. *nana*, which can be planted in 6–12in (15–30cm) of water. Even so, its blooms, at about 2in (5cm), are twice the size of those of the smallest hardy hybrid, *N. pygmaea* 'Alba'. The most popular tropical lily is the medium-to-large, day-blooming, scented hybrid, *N.* 'Blue Beauty'.

Lily Planting Depths

With so many species, hybrids and cultivars of lily available, planting depths vary enormously from one end of the lily "spectrum" to the other. Perhaps the best way of determining the planting depth for your chosen lily is to check the category under which the selected plant is being sold.

Most lilies are sold with this information, along with details relating to correct positioning, colour of blooms and recommended planting depth. As a rough guide, the following planting depths may be used where this information is not readily available at purchase.

Category	approx. suggested planting depth
Miniature/ Pygmy/Dwarf	10–30cm (c.4–12in)
Small	30–45cm (c.12–18in)
Medium	45–90cm (c.18–36in)
Large	90–120cm (c.36–48in)
Extra large/vigorous	120cm upwards (c.48in upwards)

NOTES
1) The majority of tropical lilies can be treated as medium, with the top end of the planting depth at around 75–80cm (c.30–32in), but there are exceptions, so, if in doubt, check this out, e.g. in appropriate literature, before purchase.
2) If a particular plant is healthy, but is not producing the expected results, check in specialist books and adjust depth accordingly.

▲ Nymphaea 'Sunrise' *is a hardy yellow variety from America with large, fragrant blooms. It should be planted in water up to 36in (90cm) deep.*

▶ *The Water Fringe or Fringed Water Lily* (Nymphoides peltata) *is a free-flowering, free-spreading hardy species whose flowers last only for one day.*

● *How are water lilies propagated?*

A few varieties of hardy lilies can be grown from seed. However, plants propagated in this way take a long time to mature. Therefore, the most commonly employed method of propagating hardy lilies is by dividing established plants during the growing season. Division may also be carried out in autumn or early spring. At such times, young shoots bearing "eyes" (growing points) are chosen and the old rootstock discarded. If desired, the eyes can be carefully removed and planted up individually, with the whole of the rootstock being discarded. Tropical lilies can usually be propagated quite easily from seed, but, again, plants propagated in this way take a long time to mature. Large tubers of tropical species and hybrids often store badly over winter, but many produce small tubers during the growing season, and these will overwinter successfully.

● *Why do the leaves of some hardy water lilies grow above the surface of the water?*

This is a sure sign that the plant in question has out-grown its container. Adequately planted lilies always have their pads lying directly on the water surface.

▲ *Several varieties of the tropical day-blooming species* Nymphaea stellata (*commonly known as the blue lotus of India, though it is not a lotus) are available.*

Water Fringe (*Nymphoides peltata*)
Also commonly referred to as the Fringed Water Lily, and scientifically as *Villarsia nymphoides*, this fast-spreading plant has leaves resembling those of a water lily. The blooms are quite different, though, and only last for a day. However, a good stand of Water Fringe will produce so many flowers that there are always likely to be at least several open each day throughout the flowering season. *N. peltata* produces roots at virtually every leaf node and so has the potential to become extremely invasive.
Habitat preference: Water down to around 36in (90cm) in full sun.
Flowering period: Summer.
Propagation: Easy; detach individual rooted sections from the long, spreading stems and plant these separately. This can be done at any time.

Golden Club (*Orontium aquaticum*)
The Golden Club can grow as a surface plant or as a marginal. Which habit is adopted depends to a large extent on the planting depth. If grown in deep water, most of the leaves will be of the surface type; in shallow water, most will be of the marginal type. The unusual, attractive inflorescences are spike-like, but without the enveloping spathes found in arums. They have greenish/brownish stalks, followed by a snow-white middle section, topped with a golden tip.
Height: Up to 18in (45cm) – see below.
Habitat preference: Water level to around 18in (45cm) for 'surface' effect; around 8in (20cm) or less for 'marginal' effect. This plant does well in full sun or partial shade.
Flowering period: Spring.
Propagation: Either from fresh (green), but fully mature, seeds sown in early summer, or by division of established plants, either in spring or autumn.

Floating Plants

LIKE SURFACE PLANTS, FLOATING PLANTS HAVE leaves that are buoyant and rest on or just above the surface; however, the roots of floating plants hang in midwater rather than anchor themselves in the substratum. Roots (or their equivalents) vary widely both in size and structure, from the single, vertical, filament-like roots of Duckweed (*Lemna* spp.) to the lush, feathery masses of gross feeders, for example *Eichhornia crassipes* (Water Hyacinth) and *Pistia stratiotes* (Water Lettuce or Nile Cabbage). Like marginals and surface species, floating plants do not oxygenate the water, but they do perform the valuable function of offering shade and protection.

The range of frost-hardy floaters is limited, but their numbers are swelled by several subtropical and tropical species that can be grown outdoors during the summer in temperate zones. These tender floaters can either be put under cover during winter or treated as annuals.

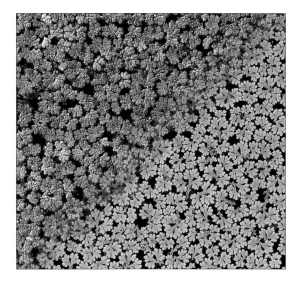

▲ *This montage demonstrates the radical change in colour that occurs between the summer (pink) and winter (green) foliage of* Azolla caroliniana.

● *Are there any special conditions attached to the buying or selling of floating plants?*

Generally speaking, floating plants are freely available everywhere. However, there are some notable exceptions in certain countries (or certain areas/ states/provinces within particular countries) regarding particularly invasive species. The Water Lettuce (*Pistia stratiotes*), Duckweed (*Lemna* spp.) and Water Chestnut (*Trapa natans*) are, for example, prohibited or restricted in at least some parts of the US. If a particular plant is openly available for purchase, it is likely to be legal. If there is any doubt, check the relevant laws for your locality, region or country.

● *Will floating plants affect conditions in the pond in any way?*

As a direct consequence of the shade cast by floaters, algae – both free-floating and filamentous types – are often controlled by being starved of light. A generous, but not excessive, covering of floating plants will also help control evaporation via the water surface and will

thus help keep temperatures down during hot spells. Gross-feeding floating species such as the Water Hyacinth (*Eichhornia crassipes*) will also consume large quantities of dissolved compounds (some of them potential polluters) and can therefore assist in keeping the water "sweet". Despite these beneficial effects, floating plants should not be allowed to cover the whole of the pond surface (see below).

● *Should floating plants be thinned out during the growing season?*

Under favourable growing conditions, some floating plants such as *Azolla*, *Eichhornia*, *Lemna*, *Pistia* and *Trapa* can spread very rapidly and cover large areas of the pond surface. In doing so, they may restrict the growth of other plants, particularly submerged oxygenators, by creating excessive shade and thus preventing oxygenators from photosynthesizing. Conditions can be made even worse if the pond also contains spreading surface plants. Thinning should be carried out as necessary during the growing season, so that at least 35–40% of the water surface is left clear of vegetation.

Selected Floating Plants

Fairy Moss (*Azolla* spp.)

There are two species of Fairy Moss used in temperate ponds: *Azolla caroliniana* and the slightly larger *A. filiculoides* (though *A. africana*, *A. nilotica* and, from Australia, *A. pinnata* could also be cultivated successfully in these areas). Despite their common name, these plants are not mosses, but floating ferns. When crowded and in shady conditions, the fronds crumple up and are pushed into lush green cushions extending a couple of centimetres above the surface and giving the plant an attractive appearance.

A. caroliniana and *A. filiculoides* can only be distinguished from each other microscopically. In *A. caroliniana*, the fine hairs that make the plant's surface unwettable are double-celled, while in *A. filiculoides*, they are single-celled. Other hairs, which are barbed and known as "glochidia", are divided by transverse walls in *A. caroliniana* and not in *A. filiculoides*. Overall, therefore, size alone is not a useful distinguishing characteristic.

Both species become bronze-red as light intensity increases during the season and both can spread rapidly into pest proportions.

Habitat preference: Full sun or partial shade.

Flowering period: Being ferns, these plants do not flower.

Propagation: The smallest fragment will reproduce, so propagation consists simply of redistributing individual plants or groups into new locations. In cold regions, it is wise to keep a small clump under cover during winter, despite the undoubted hardiness of both species.

Water Hyacinth (*Eichhornia crassipes*)

Water Hyacinths have become troublesome pests in some tropical and subtropical rivers and waterways. However, in ponds, this elegant plant can be easily controlled. In temperate zones, plants will die over winter, unless offered protection under cover, but in subtropical and tropical areas, they will survive quite happily throughout the year.

In temperate regions, Water Hyacinths are grown in ponds during the summer months primarily for their bulbous-based floating leaves, since flowering is rare under such conditions. Elsewhere, they are also cultivated for their beautiful blooms. Additionally, their lush, feathery roots play an important water-purifying role if sufficient plants are growth together.

Height: Up to 24in (60cm) or more in the wild, but invariably about half this size in ponds.

Habitat preference: Still or gently flowing waters in full sun.

Flowering period: Mainly summer, though occasional plants have been seen flowering in the wild as late as November.

Propagation: Mature plants produce plantlets or runners around their periphery and these can be gently prised off or cut off. Plants may be overwintered in water in well-lit quarters. Mature specimens may also be planted in a sand/peat mixture and kept moist under well-lit conditions at temperatures of around 15°C (59°F) or slightly higher.

Frogbits (*Hydrocharis morsus-ranae* and *Limnobium laevigatum*)

The temperate Frogbit (*Hydrocharis morsus-ranae*) has been a favourite of European pond keepers and water gardeners from the earliest days of the hobby. It has small lily-like leaves arranged in rosettes and produces delicate three-petalled white flowers with vivid yellow centres.

A beautiful tropical representative from the same family (the Hydrocharitaceae) has long been available for tropical aquaria, is grown in ponds in tropical regions, and could, potentially, be introduced into temperate ponds as a temporary summer addition. Amazon Frogbit (*Limnobium laevigatum*) is a fast spreader in ideal conditions.

▲ *The Water Hyacinth* (Eichhornia crassipes) *only rarely flowers in temperate regions, where it is grown primarily for its unusual foliage.*

Habitat preference: Still waters in either full sun or in partial shade.

Flowering period: Summer.

Propagation: In *H. morsus-ranae*, individual rooted plantlets can be separated at any time during the growing season. Winter buds (turions) can also be collected in late autumn and kept in a submerged container in the pond itself or in alternative cool quarters until the following spring. In *L. laevigatum*, individual rooted plantlets (produced on runners) can be separated at any time during the growing season. In temperate regions, the Amazon Frogbit needs to be overwintered indoors under good lighting and warm temperatures.

Duckweeds (*Lemna* spp.)

Duckweeds (*Lemna* spp.) are nearly always introduced into ponds by accident. They are highly invasive plants that are very difficult to eradicate and so tend to be avoided by pond keepers and water gardeners. This is a pity, as all the species are attractive plants when seen close-up.

Of the three species that are found in temperate regions – *Lemna gibba* (Thick Duckweed), *L. minor* (Lesser Duckweed) and *L. trisulca* (Ivy-leaved Duckweed) – the first two are generally regarded as weeds,

▲ *Water Lettuces* (Pistia stratiotes), *seen here beyond the water lilies, are attractive and fast-spreading under the appropriate conditions.*

while *L. trisulca* is somewhat less invasive. Other Duckweeds that are also occasionally encountered are *Spirodela* (*Lemna*) *polyrrhiza* and the more tropical (and smaller) *Wolffia arrhiza* (Dwarf Duckweed).

Habitat preference: Still waters in full sun.

Flowering period: Summer, but the tiny flowers are difficult to spot.

Propagation: Even individual single-leaved plants will reproduce rapidly.

Water Lettuce (*Pistia stratiotes*)

Widely distributed throughout tropical and subtropical regions, where it is sometimes regarded as a pest, the Water Lettuce or Nile Cabbage is a fast-spreading floating plant with downy/velvety rosettes and feathery roots. It is an excellent surface-covering plant, which can be cultivated outdoors during the summer in temperate regions, but requires winter protection. Elsewhere, it can be left in ponds throughout the year, although, at temperatures below 15–20°C (59–68°F), it begins to regress, particularly if the length of daylight is below 10–12 hours.

Height: Up to 12in (30cm) reported, but this is exceptional; most plants remain below 6in (15cm).

Habitat preference: Still waters in full sun.

Flowering period: Summer, but the blooms are inconspicuous and very often remain undetected.

Propagation: Individual plantlets (produced on runners) may be separated from the parent plant at any time during the growing season. Tiny plants, which arise from undetected seeds, are occasionally found on the surface.

Salvinia (*Salvinia* spp.)

Several species of *Salvinia* are available, either as permanent occupants of tropical and subtropical ponds, or as summer floating plants in temperate regions (but see below).

The most commonly encountered species is the highly variable *S. auriculata* (Butterfly Fern). Each plant consists of two floating fronds (leaves) and a third, finely divided frond that hangs vertically below them and looks and acts as a root. However, as plants are joined together by spreading lateral shoots, they create the impression of consisting of larger, multi-leaved plants.

A hardier species that can be kept longer in temperate zones (even from year to year in milder regions) is *S. natans*, which readily forms spores before disappearing or dying back in autumn.

Habitat preference: Still waters in full sun.

Flowering period: These plants are ferns and therefore do not flower.

Propagation: Separate individual plants during the growing season. Fruiting bodies ("sporangia") may also be collected from *S. natans* (*S. auriculata* rarely

▲ *Duckweed* (Lemna *spp.*) *growing among the pineapple-like leaves of Water Soldier* (Stratiotes aloides). *Under conditions such as these, Duckweed is extremely difficult to keep under control.*

produces sporangia) in late summer or autumn and stored in cool, but frost-free, conditions until ready for planting the following season.

Water Soldier (*Stratiotes aloides*)

The Water Soldier is a dioecious plant, i.e. the plants have either male or female flowers. It also exhibits an interesting distribution pattern in the United Kingdom, where plants found in northern regions are entirely female, while those in the south are predominantly male. Flowers of both types occur only in an intermediate zone between the northern and southern populations. Occasionally, hermaphrodite flowers, i.e. blooms carrying male and female parts, are also encountered.

S. *aloides* is also interesting in that it sinks to the bottom of the pond in winter and rises during the growing season. This "behaviour" is thought to be linked to the varying calcium carbonate content of the leaves as the plant goes through its yearly cycle.
Height: Up to 20in (50cm), but usually smaller.
Habitat preference: Still, calcium-containing (hard) waters, in full sun.
Flowering period: Summer.
Propagation: This is usually achieved by separating offsets from the parent plant during the growing season. This species also frequently overwinters as a small sunken plantlet or bud.

Water Chestnut (*Trapa natans*)

This rosetted floating plant is grown in ornamental ponds predominantly for its attractive foliage (though its fruits are edible). Unlike the vast majority of pond plants, the Water Chestnut is an annual, and thus needs to be regrown from its nut-like fruits each year.
Habitat preference: Warm, still conditions in full sun.
Flowering period: Summer.
Propagation: Ripe fruits are collected at the end of the season and stored in cold water over winter until they re-sprout the following spring. If the water in the pond is not excessively shallow, the fruits usually survive temperate winters outdoors.

Submerged Plants

SUBMERGED PLANTS – WHICH ARE ALSO KNOWN as oxygenators – are the only plants that can directly influence the levels of dissolved carbon dioxide and oxygen in a pond through photo-synthesis. Many species also provide food, shelter and spawning sites for fish, as well as acting as a buffer against fluctuations in water quality. Oxygenators are therefore vital members of the pond plant community.

In temperate regions, the selection of species available is not extensive, but, as we move into warmer areas, the range becomes more varied until, in tropical regions, most of the plants that are generally sold for aquaria can be considered for outdoor cultivation. Details of such species may be found in specialist aquarium texts (see Further Reading, page 202).

Just as the distinction between bog and marginal plants, or between shallow-water and deep-water marginals is unclear, so is that between submerged, marginal and surface plants, with some so-called submerged types producing aerial or surface leaves and/or inflorescences at certain stages in their life cycle.

● *Should oxygenators be grown in containers or allowed to root freely along the pond bottom?*

There is no hard-and-fast rule. If planted in containers, then – at least initially – they are easier to control. However, once they begin to grow, many oxygenators spread out of their containers and swiftly root themselves along the bottom of the pond. Unlike marginals and vigorous surface plants, such as water lilies, oxygenators can easily be thinned out.

● *Are oxygenators a threat to pond inhabitants?*

Only bladderworts – and then only to newly-hatched fry. Bladderworts have finely divided underwater foliage but no roots, and tiny bladders with which they trap small creatures such as water fleas (*Daphnia*).

● *Can oxygenators unbalance conditions in a pond?*

During the day, oxygenators absorb carbon dioxide from the water and release oxygen. This reduction in carbon dioxide makes the water less acidic (higher pH). When photosynthesis stops at night, the water regains acidity. These fluctuations pose no problem as long as a pond is not overstocked with oxygenators. If it is, the more sensitive species of fish may be affected.

● *How many submerged plants should I use?*

Submerged plants should be stocked at the following levels: for small ponds, about 20 plants per square foot (approx. 900sq cm); for medium-sized ponds, use about 10 plants over the same area; for large ponds, reduce this to 5–6.

▲ *Close-up of Hornwort* (Ceratophyllum *spp.*) *producing oxygen bubbles. These are tough, wiry plants whose "roots" consist of modified leaves.*

Selected Submerged Plants

Starworts (*Callitriche* spp.)

Starwort species are difficult to identify, since their leaf shape and overall plant appearance is greatly influenced by ambient conditions. All types have light green, delicate-looking underwater foliage that spreads out as floating rosettes on reaching the water surface.

The species most widely available are *Callitriche autumnalis* (which is, almost certainly, *C. hermaphroditica*) and *C. verna* or *vernalis* (which is almost certain to be *C. platycarpa*).

Habitat preference: Water down to around 18in (45cm) in full sun.

Flowering period: Summer, but the flowers are inconspicuous.

Propagation: Divide established plants during the growing season.

Hornworts (*Ceratophyllum* spp.)

There are two very similar species of Hornwort: *C. demersum* and *C. submersum*, the latter of which appears better suited to warmer temperatures than the former. On close examination, *C. demersum* – the type usually available for ponds in temperate zones – has tougher, darker leaves than *C. submersum* and possesses fewer leaf tips (2–4, instead of 5–8).

Hornworts are unusual in that they do not produce roots. However, where the stems come into contact with the substratum, modified leaves are formed that help anchor the plants.

Habitat preference: Any depth, from a few centimetres to well over 39in (c.1m), in full sun or partial shade.

Flowering period: Summer, but the flowers are inconspicuous.

Propagation: Divide established plants during the growing season. Turions (winter buds) are also produced, and are easily dispersed by flowing water.

Pondweeds (*Egeria, Elodea, Lagarosiphon*)

These three oxygenators are often confused with one another. There are, however, some distinguishing characteristics between the three most commonly available representatives of these genera.

Egeria densa

'Densa', Argentine Pondweed, Giant Elodea

Light-coloured leaves around 1–1.2in (2.5–3cm) in length, occasionally up to 1.6in (4cm); up to 0.2in (4mm) wide; leaves occasionally lightly twisted; largest of the Elodeas.

Elodea canadensis

Canadian Pondweed, Elodea

Somewhat darker than *E. densa*; leaves only around 0.4in (1cm) long and 0.1in (3mm) wide; leaves curved slightly downwards.

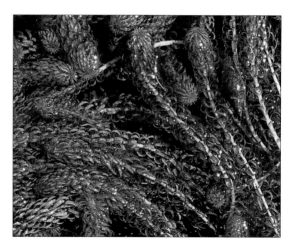

▲ *Elodea 'Crispa'* (Lagarosiphon major) *is extremely easy to establish in ponds. It tends to become invasive, so you should thin the plant out regularly.*

Lagarosiphon major

'Crispa', African Elodea

Dark green leaves around 1.2in (3cm) long and 0.1in (3mm) wide; leaves are strongly recurved and often covered in "lime" deposits.

Habitat preference: Deep water in full sun.

Flowering period: Summer, but the small flowers usually go unnoticed.

Propagation: Stem cuttings may be taken at any time during the growing season, bunched together with a lead strip and allowed to sink to the bottom of the pond, or planted in containers.

Hair Grasses (*Eleocharis* spp.)

Hair Grass is usually sold as an underwater plant for aquaria, but a number of these species are also available for planting in ponds. The most popular of these dual-purpose species are *Eleocharis acicularis*, *E. minima* and *E. vivipara*.

Grown as submerged plants, the delicate beauty of these species tends to be lost among the other, more robust and broader-leaved pond oxygenators. However, when the Hair Grass species are cultivated in shallow conditions, the leaves will project through the water surface and produce small, unspectacular inflorescences.

Hair Grass species can also be grown as bog plants.

Habitat preference: Moist soil, down to around 24in (60cm) depth, in full sun.

Flowering period: Summer.

Propagation: Divide established plants during the growing season. In *E. vivipara*, the characteristic plantlets produced in tiers along growing shoots can be separated and replanted.

Willow Moss/Java Moss
(*Fontinalis* and *Vesicularia*)
Willow Moss (*Fontinalis antipyretica*) is usually
found in flowing streams in temperate zones, but will
adapt to conditions in most ponds. In tropical areas,
the narrower-stemmed *Vesicularia dubyana* (Java
Moss) forms similarly dense clumps. Both plants are
beautiful when seen at close quarters in shallow water.
Habitat preference: Shallow to deep water in full sun
or partial shade. *F. antipyretica*, in particular, does
well in flowing water around streams and cascades.
Flowering period: These species are mosses and therefore do not flower.
Propagation: Clumps may be separated during the
growing season, and, if possible, should be attached
to a base (e.g. a rock) with an elastic band until they
anchor themselves. Pre-attached clumps of *V.
dubyana* are widely available for aquarium use, and
will transfer safely to tropical ponds or to temperate
zones during warm summers.

Water Violet (*Hottonia palustris*)
The Water Violet, like Hair Grass, can produce
underwater and aerial foliage. The finely-divided submerged leaves are efficient oxygenators. The emersed
leaves are also divided but do not oxygenate the
water. The white to lilac flowers form spikes that are

▲ *A Water Violet* (Hottonia palustris) *flower spike
surrounded by white and gold flowerheads of Golden
Club* (Orontium aquaticum).

held well clear of the water surface. Together with the
aerial leaves, they give the Water Violet the appearance of a marginal.
Height: Flower spikes around 12–18in (30–45cm) or
taller in ideal conditions.
Habitat preference: Shallow to deep water up to 24in
(60cm) in full sun or partial shade. Acid water
appears to be preferred.
Flowering period: Early to mid-summer.
Propagation: As the season ends, fragments break off
and form winter buds (turions) which can be replanted for the following season. Alternatively (and more
practically), underwater stem cuttings may be taken,
or clumps divided, during the growing season.

Milfoils (*Myriophyllum* spp.)
There are about 40 species of *Myriophyllum*, many
of which are suitable either for aquarium or pond
cultivation in temperate zones. A considerable number of aquarium species are also able to grow in
ponds in tropical regions.
   All are typified by very finely divided leaves which,
in some species, project above the water. This occurs
in the case of *M. proserpinacoides* (Parrot's Feather),
which was formerly known as *M. brasiliense* and is
still often referred to as such. Two other species that
are commonly available for temperate ponds are
M. spicatum (Spiked Milfoil) and *M. verticillatum*
(Whorled Milfoil). The choice for tropical schemes is
much wider.
Height: Around 12in (30cm) or more for *M. proserpinacoides* and *M. verticillatum* grown in shallow
water; others produce predominantly, or exclusively,
submerged leaves.
Habitat preference: Moist habitat, down to 24–36in
(60–90cm) depth in full sun, depending on species.
Flowering period: Summer, but flowers are usually
small or inconspicuous.
Propagation: Take stem cuttings at any time during
the growing season. Alternatively, self-rooting stems
are produced by some species and these may be separated and replanted.

Curled Pondweed (*Potamogeton crispus*)
This attractive oxygenator has wavy green/bronze-
coloured submerged leaves. It is a fast spreader, so
may require regular thinning during the growing
season. It is widely distributed in the northern hemisphere, where it can colonize large bodies of open
water. The stems and leaves are a little brittle and
have the overall appearance of a marine macro-alga
(seaweed), rather than a flowering plant. *P. crispus*
is a tough, frost- and ice-resistant species that can
be grown even in shallow ponds in temperate zones.
Habitat preference: Water down to 36in (90cm) in
depth, in full sun.

Flowering period: Summer, but the small flower spikes – standing at only c.1.2in (3cm) – are inconspicuous.

Propagation: This is easily achieved by separating free-rooting stem cuttings from the parent plant at any time during the growing season.

Water Crowfoot (*Ranunculus aquatilis*)
This is a widely distributed plant that is found in temperate and slightly warmer regions in both Europe and the Americas. Its finely divided submerged foliage bears a superficial resemblance to that of *Myriophyllum*, but in *R. aquatilis*, the finely divided leaves are palm-like and have a stalk, rather than being arranged in whorls around the central stem. Water Crowfoot forms small clover-like floating leaves, from which the five-petalled white flowers with their bright yellow centres arise.

Habitat preference: Water down to around 39in (c.1m) deep, in full sun.

Flowering period: Summer.

Propagation: Stem cuttings can be bunched together with a lead strip and allowed to sink to the bottom of the pond, or planted in containers. Alternatively, freshly-collected seeds may be sown in late summer.

▲ *Some Milfoils* (Myriophyllum *species*) *produce both submerged and aerial leaves, which make them appear to be marginals rather than oxygenators.*

Other Oxygenators

Submerged plants that have traditionally been regarded as aquarium subjects are gradually beginning to feature as additional or alternative choices for ponds. In general terms, any oxygenator can be considered for pond culture as long as the conditions are appropriate. However, many types are too short for their beauty to be appreciated in deep ponds or in those already well stocked with plants. Thus, most species of Eel or Tape Grass (*Vallisneria* spp.) are unlikely to be noticed. One notable exception is *V. gigantea*.

Many tropical oxygenators, like their temperate counterparts, exhibit different growth forms, some producing both submerged and aerial leaves and blooms. Others, like the tender *Ceratopteris thalictroides* (Indian Fern), are so versatile that they can grow as bog plants, submerged oxygenators, or even floaters. Details of such plants may be found either in general aquarium texts or in specialist plant publications (see Further Reading, page 202).

Pond Wildlife

IRRESPECTIVE OF WHERE A POND IS SITUATED –
whether deep in the countryside, or in the heart
of a city – it will, sooner or later, be visited by a
range of animals that are not part of the community
originally introduced into it. These arrivals are regarded
as welcome or otherwise, and can become temporary or
permanent residents, or regular or infrequent visitors,
depending on circumstances and the pond owner's wishes.

As well as playing host to these uninvited guests, the pond
keeper may plan to introduce certain native species of animals
(or encourage them to introduce themselves). The aim of such
an exercise is to establish a community of local fauna, or of species
that, although not necessarily native to the region, are nevertheless
wild-type representatives of their kind.

In many cases, visiting or introduced wildlife can live quite happily
alongside resident ornamental fish, either co-existing without paying
much attention to each other, or simply avoiding each other. Frogs
and toads, for example, generally fall within these categories.

Other types of wildlife, however, pose a very real threat to the
pond inhabitants, whether these be ornamental or native fish, or
other forms of wildlife. Perhaps the best known examples of
such unwelcome wildlife visitors are birds that prey on fish
and amphibians, notably herons.

The range of wildlife species that are associated with ponds
is extensive, and varies widely from country to country
and region to region. This section provides an overview
of the main groups of animals that are generally
encountered, in addition to highlighting
a number of selected examples.

▶ *Common Frog* (Rana temporaria) *resting in
a pond full of Duckweed* (Lemna *sp.*).

Native Fish

NATIVE OR WILD-TYPE FISH HARDLY EVER arrive in a pond of their own accord, other than in systems that are fed by streams or other watercourses. Though many incidents have been reported of fish appearing in ponds from no-where, they are usually attributable to eggs being introduced accidentally with plants that have been bought for ornamental, food or spawning purposes.

What is a Native Fish?

Strictly speaking, the term "native fish" denotes species that are indigenous to a particular area. However, the term has become more loosely applied, and this has led to some confusion.

For example, the American Flagfish (*Jordanella floridae*) is predominantly native to Florida, but not to, say, California. Yet, in the latter state, it can also be referred to as native, in the wider context of being a native American species.

Another, more extreme, example is the Eastern Mosquito Fish (*Gambusia holbrooki*). Its native home extends from Central Alabama, eastwards to Florida, and northwards along the Atlantic seaboard to New Jersey. While this natural range is far wider than that of the Flagfish, it certainly does not encompass the far-flung parts of the world where the species has been introduced for malarial control. Yet anyone unaware of this would consider a Mosquito Fish collected from a drainage ditch in Spain or Malaysia as a *bona fide* "native" species.

The term "native" is also often applied to the wild type of any species, irrespective of whether or not it is indigenous to the area or country. The best example of this is the olive-green/brown wild form of the Common Goldfish (*Carassius auratus*), which is regarded as being the "native" form of the species, while the ornamental, col-oured varieties are deemed "cultured". In these cases, though, the label "native" is omitted, since it refers to coloration rather than origin.

▲ *The American Flagfish* (Jordanella floridae) *was first imported as a "coldwater", rather than a tropical fish, but beware – it cannot survive temperate winters outdoors.*

● *Does keeping native fish in ponds contribute to their conservation?*

... This depends on the species concerned and on the approach adopted, or allowed by law. In the majority of cases, native fish that are freely available for keeping in ponds are not endangered. Many such stocks are, in fact, captive-bred in origin, rather than wild-caught, so – in these instances – there are no conservation implications. In cases where a species is known to be under threat of some kind, availability (if there is any) is generally strictly controlled. As a rule, such fish are only accessible to members of specialist societies, or those of a registered group dedicated to the conservation of the species. In these instances, important conservation work can undoubtedly be achieved in ponds, but only when it is carried out as part of a well-coordinated captive breeding programme.

● *Will native fish interbreed with ornamentals?*

If fish are biologically closely related to each other, they will interbreed, irrespective of whether they are native or ornamental in origin. For example, Goldfish of any variety will breed with the wild type, as will Koi with Common Carp, etc. Therefore, if pure lines are to be maintained, native fish must not be housed in the same pond as cultured varieties of either their own, or closely related, species.

Stocking With Native Fish

Native fish can be accommodated in ponds alongside ornamental fish, as long as there is no danger of predation. Similarly, if you are establishing a collection of strictly local species, then you must also ensure that these are compatible with each other.

Ponds that are exclusively reserved for local species are usually set up and managed on a more scientific basis than those housing mixed selections, in so far as conditions in the former are generally more closely matched to the (often very specific) needs of the species in question. In "locally native" schemes, greater attention is also paid to the stocking level of each species, reflecting their relative abundances in wild populations.

Sourcing suitable stocks of native fish can be problematic. The generally less spectacular coloration of native fish means that they are not as much in demand as exotic ornamentals, and so are not stocked as regularly by most aquatic retailers. There are, however, a few outlets that dedicate a small section of their sales area to native species, and these are well worth the effort of seeking out. In addition, some members of local pond-keeping societies who are interested in native species may breed them. Generally speaking, fish obtained from these sources are of excellent quality.

The same cannot be said of specimens collected from the wild, which may harbour parasites and other pathogens. These fish may well be able to tolerate their pathogen load quite easily and safely in the wild, but the stresses of capture will weaken their resistance. As a result, wild-caught fish frequently develop some form of infection after capture. Even if they don't, the fact that they are carrying such pathogens places the resident pond population at risk, unless adequate quarantine and acclimatization procedures are adopted.

Exciting though it is to collect fish from the wild, you should never do so without first checking its legality. In some countries, the activity may be legal for all species except those designated as endangered or under threat. In others, collection may be severely restricted or prohibited, while in yet others, the legislation on the collection, sale and captive rearing of native fish may vary between different states or provinces (see Wildlife and the Law, pages 178–179).

So, as long as no laws are being contravened, and provided the selection and sources of native species are appropriate, there is no reason why native or wild fish should not be kept in garden ponds. Many people are coming to realize that colour isn't everything, and opting for a more natural aspect to their ponds by keeping locally-occurring species instead of, or as well as, ornamentals. The two groups may be kept together or separately, depending on compatibility.

▶ *An obviously incompatible combination of native species! A predatory Pike* (Esox lucius) *and its intended victim, a Nine-spined Stickleback* (Pungitius pungitius).

Native Amphibians

UNLIKE FISH, AMPHIBIANS CAN JOURNEY FOR long distances overland, and can thus colonize ponds that are nowhere near a natural source of either eggs or adult animals. However, in many parts of the world, growing urbanization is exerting increasing pressure on land and natural resources, and amphibians – in common with other creatures – have suffered badly as a result. Large numbers of ponds have either disappeared or become far too polluted to support life; where there are such obstacles to natural colonization, other means must therefore be sought.

In recent years, politicians, administrators and commercial interests have begun to show greater ecological awareness when planning and implementing change. As a consequence, the need for continuing economic growth is now tempered with a degree of concern for the environment, including wildlife. This trend is particularly prevalent in the USA and northern Europe. In several countries, for example, the mass slaughter of amphibian populations on busy roads during spring is being tackled by providing "tunnels" for toads and frogs to migrate to and from their spawning sites. Many of these programmes are proving so successful that declines in local amphibian populations have been arrested or even reversed.

The proliferation of garden ponds has been a major contributory factor to amphibian conservation. They provide migrating frogs and toads, as well as salamanders and newts, with badly needed havens. One of the many attractions of these garden refuges is that, once adult amphibians visit a pond and breed in it, continuity over succeeding years is virtually assured.

Stocking with Native Amphibians

While many ponds become colonized by amphibians as a matter of course, others need a little help. This can be provided in several ways.

Introducing adult frogs or toads once the spawning period has passed is not a particularly efficient method of establishing a resident population. These adults have bred elsewhere and are therefore much more likely to wander off in search of their "home" pond than individuals which have either moved in naturally, or have actually grown in the ponds. There is, however, a direct correlation between the degree of success achieved in introducing adults and the size of the garden surrounding the pond. Frogs and toads are more likely to remain in larger sites than in smaller ones.

A far more successful way of establishing a resident frog or toad population is to obtain spawn in the spring from a safe (and legal) source. Froglets arising from such spawns identify the pond in which they have developed as home and will therefore be more likely to return the following spring. It takes frogs and toads

● *Do amphibians pose any threat to pond fish?*

**... It is sometimes claimed that male frogs and toads pose a threat to fish during the breeding season, through their habit of grabbing at any moving object in their urge to secure a mate. Slow-moving fish, such as some of the fancier varieties of Goldfish, may be caught by amorous male frogs and toads. However, this is rare, and the fish is usually released as soon as the amphibian realizes its error. The only potentially hazardous situation for Fancy Goldfish is when the male amphibian population is so high that frequent mistakes are likely to occur. Removing the Goldfish to temporary quarters at the peak of the frog/toad breeding season may be prudent, if not strictly necessary. Fish fry are regarded as food by newt and salamander young, but they are more likely to fall prey to the adults of their own kind.

● *What is the difference between a frog and a toad?*

So-called true toads belong to the family Bufonidae, which is characterized by a predominantly terrestrial habit; dry, warty skin (often capable of secreting toxic compounds); and short "walking", rather than "jumping", legs with relatively weak webbing between the toes. Frogs can belong to a number of families, the best-known of which is the Ranidae. Frogs have a generally smoother skin, and are more aquatic in habits, than toads. They also possess a thick covering of slippery mucus (which is waxy in some types) and "jumping", rather than "walking", legs. However, there are exceptions, and some frogs, e.g. *Xenopus* spp. from the family Pipidae, are often referred to as toads.

● *How do frogs and toads avoid interbreeding with each other?*

In temperate zones, the breeding season for frogs generally starts earlier in the year than for toads. This overcomes at least some of the potential risks of hybridization. Moreover, even during the period of overlap, toads tend to choose deeper water than frogs. Yet, despite these natural hurdles, there are times when a male toad will grab a passing frog, and vice-versa. Usually, when this happens, the grabbed individual croaks, causing the "grabber" to realize his mistake and release his intended mate. In addition, there are genetic incompatibility barriers between frog/toad eggs and sperm.

▼ *Mass spawning of Common Frogs* (Rana temporaria). *Frogs and toads remember the pond they developed in, and will attempt to return to it for spawning.*

several years to attain full maturity, so stocking with spawn for the first two springs is advisable.

With newts, introducing spawn is an effective method of establishing a colony. However, newt eggs are laid singly, rather than in masses (frogs) or in strings (toads), which makes the exercise somewhat more difficult. Fortunately, newts do not appear to possess such highly developed wandering instincts as frogs and toads. Therefore, a better way of establishing a colony of these amphibians is to introduce a small number of adults to the pond in early spring, once they have entered their aquatic (breeding) phase. Unlike frogs and toads, newts have a long breeding season and thus remain completely water-bound for a period of many weeks. The duration of this aquatic phase varies according to the species and its location.

In the case of salamanders, most are predominantly terrestrial, and return to water only to breed. A notable exception is the Axolotl (*Ambystoma mexicanum*), which spends its whole life in water and retains the external gills – a characteristic of the larval stages of all newts and salamanders – throughout. Many species of salamander lay eggs, but others, such as the very striking European Fire Salamander (*Salamandra salamandra*), are livebearers.

Establishing a salamander population around a garden pond is considerably more difficult to achieve than with newts. The best way of doing so is to introduce either spawn or larvae to a pond and allow these to develop to maturity (once they leave the water) in a spacious garden.

Caring for Native Amphibians

Adult frogs spend much of their time after spawning in and around water. Suitable resting and sunning places should therefore be provided for them. Toads are considerably more terrestrial and so will tend to move further away from the pond once spawning is over; the same is true of salamanders. A variety of accessible shelters in the form of logs or rocks should be

▼ *Red-spotted Newts* (Notophthalmus viridescens) *in courtship. Afterwards, the male drags the female over to his spermatophore (sperm packet) for fertilization.*

provided for these creatures, both in the immediate vicinity of the pond, and in shady locations further afield. These shelters will also be used by newts once they leave the water.

Young amphibians, particularly frog tadpoles, are preyed upon by a wide range of animals, including dragonfly nymphs, fish and birds (see Other Wildlife, pages 172–179). Where possible, these predators should be kept under control, but, even in their presence, many young amphibians will survive through to metamorphosis (i.e. to the stage where they change from aquatic larvae into their final terrestrial form), so long as the pond contains a good stock of submerged vegetation.

During their aquatic phase, frog and toad tadpoles, efts (newt larvae) and young salamanders will usually find adequate supplies of food within a well-established pond. If, however, a shortage is suspected, flaked or pelleted fish foods, plus a regular supply of live foods such as waterfleas (*Daphnia*) can be offered.

During metamorphosis, easy exit routes should be created for the emerging young. Several methods of doing this have already been detailed in Pond Set-up.

Frogs and toads tend to emerge in a continuous, protracted flow that lasts for several weeks. During this time, they can fall victim to a wide range of predators, not only the obvious ones, such as cats and herons, but also more unlikely ones, such as blackbirds and thrushes. It is thus advisable to provide the newly metamorphosed young with some form of protection. The simplest way of doing this is to allow pondside vegetation to grow unchecked for about a month and to refrain from mowing lawns for as long as possible (several weeks at least) when emergence is at its peak.

However, if you decide that you must mow your lawn during this critical period, the most humane course of action is to inspect small areas of turf in advance, and transfer any young frogs or toads to safety. This is a laborious process, but it does ensure your amphibians a fighting chance of survival. After all, their lives are full of enough hazard without the added danger of being trodden underfoot or cut to pieces by a mower for the sake of a manicured lawn.

▲ *A larval Fire Salamander* (Salamandra salamandra). *Salamanders may take several years to reach maturity, so annual stocking with larvae will help the population.*

● *What is the difference between a newt and a salamander?*

... Newts and salamanders are closely related biologically; most species and genera belong to the family Salamandridae. The dividing line is thus impossible to pinpoint. Traditionally, the more aquatic species are known as newts, while the more terrestrial are referred to as salamanders.

● *Should hibernating amphibians be handled?*

Amphibians, like fish, are poikilothermic, i.e. their body heat fluctuates in accordance with ambient temperature. During hibernation, their metabolism slows down dramatically until after the winter, when it begins to speed up once more as temperatures rise. During this dormant or semi-dormant period, the small amount of energy that is used comes from stored fat reserves laid down during the previous autumn. If a hibernating amphibian comes into contact with a heat source, such as the warmth given off by human hands, it reacts by raising its metabolic rate. If the handling is prolonged, there is a very real risk of the amphibian beginning to reawake. If it is subsequently put back into cold surroundings, it will once again have to slow its metabolism down and re-enter a state of hibernation, but this time with somewhat lower reserves. Such disruption does not do the amphibian any good, and may even do a great deal of harm. Therefore, avoid handling hibernating amphibians. If moving a hibernating animal is unavoidable, do it as quickly as possible, keeping contact to an absolute minimum and gently lifting up the amphibian with a spoon, ladle, trowel or other implement that is at the same temperature as the surroundings.

Native Reptiles

IN SOME COUNTRIES, REPTILES AS LARGE AS crocodiles and alligators may visit ponds situated either in rural areas or close to natural waterways. When and if this happens, the relevant authorities should be contacted without delay, not just because these reptiles are dangerous and difficult to handle, but also because they may be protected and trapping them, or harming them in any way, may be illegal.

Snakes, too, may visit ponds, particularly in tropical and subtropical countries, though this can also happen occasionally in temperate regions. Unless you are absolutely certain that the species in question is neither dangerous nor venomous, you should not attempt to handle it, but should seek expert advice from an official department, or an experienced herpetologist.

It is, of course, perfectly possible for snakes and humans to co-exist and, where there is no

Q&A... ● Can terrapins be collected from the wild and released into garden ponds?

In many countries, terrapins are protected and cannot be collected from the wild. However, captive-bred specimens are often available. In countries such as the USA, in response to recent controversy over hygiene and small children, these stocks are usually advertised as being uncontaminated by the Salmonella bacterium. In others, imports of some exotic species that are suspected or known to pose a threat to native species are either restricted or banned. In such instances, collection of endemic species from the wild is usually prohibited.

● What should terrapins be fed on?

Terrapins will eat a wide range of predominantly meat- and fish-based foods. In the wild, they are active predators and often feed on dead animals as well. For specimens kept in ponds, anything from earthworms to pieces of fish and meat, along with any of the deep-frozen or freeze-dried chunky foods, will be accepted. So will many of the fish pond pellets, particularly those designed for trout and other carnivorous species.

threat to either party, this option is worth considering. However, although this idea may be appealing, bear in mind that many species will eat fish, amphibians and other creatures. You should also be aware that many people may be uneasy about having a snake in their neighbourhood. If we create a water scheme that attracts these creatures to a locality where they would normally not exist, it could be argued that it is our responsibility to ensure that other people do not have to live with the consequences of our decision, or are put at any risk by it.

The only other reptiles that are likely either to visit a pond, or to be considered as candidates for stocking, are terrapins (also known as turtles or sliders). Like snakes, these will prey on fish and amphibians. However, if there are only a few of them, and the pond is large, with a sizable fish and amphibian population, it may be possible to maintain a reasonable balance between the reptiles and the other pond inhabitants. Otherwise, removal to a safe, alternative site may be necessary to protect the resident pond community (but check on the legality of both this action and the relocation of the species).

Terrapins are messy eaters and produce copious amounts of waste. Some enthusiasts therefore prefer to remove them from the pond at feeding time, placing them either in a large bowl or a small, bare, prefabricated pond set aside for this purpose, until they have eaten and (preferably) defecated. Alternatively, terrapins may be fed in the pond, as long as the water treatment system can cope with the waste or the pond itself is large enough to process the increased biological load. The more animals, and the larger their size, the greater the pollution problem – thick, "soupy" green water being one of the earliest signs that the system cannot handle the waste.

▶ *The best-known terrapin species is the Red-eared Slider* (Trachemys scripta elegans). *Like other freshwater terrapins, it needs a dry spot for basking.*

Other Wildlife

▲ *A dragonfly* (Aeshna cyanea) *resting on the nymphal case from which it has emerged. Adults patrol the water surface but pose no threat to aquatic life forms.*

AS WELL AS FISH, AMPHIBIANS AND REPTILES, ponds attract a wide variety of other wildlife. Many of these species are small and often difficult to spot from a distance. Others are much larger, but frequently go unnoticed, owing to the fleeting nature and time of their visits. Small insects are good examples of the former category, while the latter includes animals as diverse as hedgehogs, raccoons, herons and foxes.

Pond Insects

Insects are among the most numerous and varied visitors and colonizers of ponds, even in heavily urbanized areas.

Some types, such as the Water Pond Skaters or Striders (*Gerris* spp.), pose no threat to other types of pondlife, except other small insects and, perhaps, the tiniest of fish fry. Skaters spend their entire lives skimming around the water surface, where they are supported by the surface tension. Their prey consists of insects that fall into the pond, although they will also feed on fish flake and floating pellets supplied for the fish.

Whirligig Beetles (e.g. *Gyrinus* spp.) and other surface dwellers are also harmless to fish and amphibians and add an interesting touch to the water/air boundaries of a pond. Their nymphs may, however, feed on small aquatic creatures.

Among the other flying insects with aquatic nymphal stages, the dragonflies and damselflies are the most spectacular. The adults pose no threat to the pond occupants, but the nymphs will prey on small fish, tadpoles and other small creatures. Bearing in mind that the nymphs of certain types like the dragonflies of the genus *Aeshna* can attain a length of 1.6in (c.4cm), the size of the prey that they can trap may be quite substantial. Caddisflies (order Trichoptera) and mayflies (order Ephemeroptera), also have predatory aquatic nymphal stages, but are no real problem, as they only eat very small creatures.

Other insects not only fly in as adults, but also spend all of their nymphal and part of their adult lives under water. Among these, the species that probably present the greatest danger to young fish, amphibians and other creatures are the nymphal and adult stages of the various diving beetles (e.g. *Dytiscus* spp.), which can grow to several centimetres in length and are strong swimmers with powerful jaws. Less of a threat, but still potentially dangerous to young fish and tadpoles, are the water boatmen, such as *Corixa*.

Pest Control

In all but a few instances, it is unnecessary to control the insect population in a pond. If the pond is appropriately stocked with fish, aquatic insects are usually kept well under control by the fish, which prey on the nymphs of all insects, including predatory ones. In a well-planted and well-stocked pond, this usually results in a balance approximating to what happens in Nature, where fairly stable populations of aquatic insect nymphs co-exist with other types of wildlife.

It sometimes becomes necessary to intervene in this natural *status quo*, by removing particularly troublesome predatory insects (usually one or other of the diving beetles). In such cases, patience and persistence are essential, since the technique is to wait for the insects to surface for air and scoop them up with a net. Thereafter, the most humane course of action is to release them into the nearest natural body of freshwater. It must be stressed, though, that, since these insects can fly, physically removing them elsewhere cannot guarantee that no future "visits" will occur. By repeatedly trying to net them, even if you fail (owing to the speed with which they can dive), you can sometimes encourage flying or aquatic predators to leave of their own accord.

Many pond keepers and water gardeners find a pond without a good insect complement quite unacceptable. Notable exceptions are those systems that have been specifically designed to display a fish collection, such as large Koi, to full effect. Making adequate provision for insects here is difficult, if not impossible, since planting the pond would detract from the main focus of attention, the Koi themselves.

Q&A...

● *Are there any other types of wildlife that are likely to visit or be found in garden ponds?*

Yes, there are, but they are far too numerous to list extensively. In addition, many are so small or secretive that they are likely to go undetected. Among these are small leeches (which usually attack invertebrates, rather than fish), flatworms, midge larvae and the like. Small though these animals are, they are gigantic compared to the incalculable numbers of microscopic and submicroscopic organisms that colonize all ponds. Among the numerous more noticeable creatures are snails which often arrive as jelly-like blobs of eggs attached to plants. Their numbers are usually kept in check by the other pond inhabitants. If a population explosion occurs, either continuous physical removal or treatment with a molluscicide may be necessary (but read all the cautionary notes regarding dosage, etc.). Crayfish may occasionally appear if the pond is close to a body of flowing water containing these creatures, but these occurrences are, at best, rare. Other crustaceans, like freshwater shrimps, may also turn up, but numbers will remain low in well-stocked ponds. In poor water conditions, the Water Hoglouse (*Asellus* sp.) and Rat-tailed Maggot (*Tubifera* sp.) may become abundant.

▲ *This Great Diving Beetle* (Dytiscus marginalis) *is feeding on a Three-spined Stickleback* (Gasterosteus aculeatus) *which it has just caught.*

Birds

All ponds attract a wide selection of birds, but the best design to cater for their needs is one that incorporates shallow areas for easy access and exit. Drinking and bathing visits, obviously, present no problems for the fish or other types of pondlife and are to be welcomed. Yet ready access also makes it easy for predatory birds to snatch fish.

Herons and many other birds that eat fish or amphibians (e.g. storks), usually land either on dry ground or shallow water and then wade into deeper water, where they will wait motionless for a passing victim. Other predatory birds, like kingfishers, select a suitable perch, such as a fence or tree, as a vantage point from where they can keep a watch out for their meal. Once this is spotted, they dive into the water, catch their prey and either return to their perch, or fly away to consume their catch elsewhere.

Kestrels and other hawks are only very occasional visitors to ponds, where they arrive in search of other birds to kill, rather than catching fish or amphibians. Their visits are fleeting, and

may never even be noticed by the pond keeper, but they create a great commotion, which can cause the fish to dash for cover and remain hidden for days or even longer, just as if they had been the intended prey.

Other bird visitors to your pond, such as ducks, are not predatory but can cause more long-term damage than the occasional heron or king-fisher. The latter may decimate your livestock, but do not affect water quality or plants in any way, where-as ducks and related wildfowl con-tinually churn up the pond substrate in their search for plants and other food. In a small or even medium-sized pond, this can wreak havoc. If they start visiting on a daily basis, their droppings will also cause the water quality to deteriorate. The resulting "green water" will prove particularly difficult to control.

▲ *Netting arranged over a pond will prove effective against predatory birds and need not look unattractive. This is a permanent set-up; others can be removed.*

Protecting Your Pond from Birds

While most bird activity around a pond is en-couraged by water gardeners, some visits must be curbed in order to safeguard the fish and other vulnerable pond residents.

Pond design can be an unobtrusive and attrac-tive method of controlling predators. However, this does not mean that you have to dispense with the shallow areas in your pond. These can remain, so long as they are protected by bird-control devices. There is a wide range available, from electrified (low-voltage) fencing to explo-sive bird-scarers. All are effective to a greater or lesser extent, but none claim to be absolutely bird-proof. Rather, their function is to scare the birds away often enough (but without harming them) that they go elsewhere in search of food.

Two points should be borne in mind with bird-control devices. Firstly, herons and other pred-ators tend to visit around daybreak. So, if you are thinking of using explosive devices, you must discuss this beforehand with your neighbours.

◀ *Even large Goldfish can be preyed upon by the Grey Heron* (Ardea cinerea), *which has a voracious appetite. Birds may be protected species; be sure to investigate.*

● *How much damage can birds cause to fish stocks ?*

... In the most extreme cases, the damage can be devastating; some pond keepers have lost all their fish. However, for this to occur, the birds need to be presented with ideal fishing opportunities and the fish must be of a suitable size for the birds to take. A pond is most at risk if it is located in an area where fish-eating birds are common, e.g. near a heronry, or if it is constructed in the flight path of such birds to and from their fishing grounds. In these cases, mechanical protection by means of a net or other suitable alternative is essential, at least during the early morning and evening when predatory activity is at its most intense.

● *Can fish be kept safely with ducks and other types of water fowl?*

The risk to fish from birds like waterfowl is not primarily that of attack (though very young fry may be eaten), but of pollution from the birds' waste. If visits from ducks and geese are regular, or if they actually form part of the resident fauna of a pond, the filtration system needs to be particularly effective. Choosing the smaller types of duck and waterfowl, and keeping them in a large, well-filtered pond stocked only with hardy species and varieties of fish, is the safest combination of birds and fish to put in your pond.

Secondly, some birds are protected by law in certain countries (e.g. herons in the US), so check in advance what control methods are permitted in your area.

Away from the shallow area, the pond surround can be designed in such a way that it does not offer any good hunting spots. Steep banks, for instance, are effective deterrents. They cannot guarantee total protection, since herons have been known to land directly in deep water, but they do help, especially in conjunction with an unobtrusive wire anti-bird fence around the pond edge. One of the best ways of discouraging kingfishers is to make sure that there are no suitable perches in the vicinity of the pond.

In severe cases, you may need to cover the entire pond with a net at night and remove it every morning. This is a labour-intensive regime, but if it is maintained for a period of several weeks, it will usually frustrate the birds into going elsewhere. To guard against attacks by kingfishers, the net may need to be kept on throughout the day during the "treatment" period. Once the danger has passed, the net can be stored and reused in autumn to protect the pond from falling leaves.

Models of raptors – hawks, eagles or owls – have occasionally been used to scare off herons and kingfishers. A model heron can also fool an avian visitor into thinking that there is already a competitor in place. Yet these decoy measures are not as successful as some of the other approaches detailed above.

Mammals

The mammals that visit ponds most frequently are hedgehogs and (although not strictly speaking "wildlife"), domestic dogs and cats. Being harmless, hedgehogs are always welcome. Dogs are more or less neutral in their effect, although they may sometimes paddle in the shallows, churning up sediment and causing some (usually reparable) minor damage in the process. Cats are almost invariably unwelcome around ponds. Less frequent, though not uncommon, visitors include foxes and, in the US, raccoons.

Hedgehogs always arrive at night. Their visits follow basically the same pattern, which may include pausing for a drink in the shallows as they forage for food. These delightful and often very tame creatures pose no threat to fish.

Cats can, with practice, become skilful fish catchers. They are also very patient, waiting motionless with their gaze fixed intently on a selected victim until it swims within range. Once this happens, all it takes is an expert swipe with extended claws and anything from a tiny minnow to a large Goldfish – or even a medium-sized Koi or Orfe – will end up either injured or, worse, scooped out onto dry land. The cat will then eat part or all of the fish, or will toy with it before leaving it to die, often under a hedge or a shrub.

Foxes, like hedgehogs, are usually night-time visitors. In general, they pause only long enough to have a drink. Occasionally, they can jump into

▲ *A cat on the prowl around a pond is a common hazard. Inset: the victim of a cat's attack, mangled but uneaten and left to die most horribly.*

● *Are there any special precautions that should be taken when handling wildlife?*

In the first instance, wildlife should not be handled, unless absolutely necessary. In the case of animals that could bite or otherwise be dangerous, such as some snakes, crocodilians, foxes, raccoons and the like, handling is best left to the experts. However, even apparently harmless species can bite, often very painfully indeed. Among these are dragonfly nymphs, water boatmen and diving beetles. Either wear protective gloves, or scoop the animals up in a net or other receptacle. If an insect needs to be picked up with bare hands, then approach it from above and grasp it around the thorax (chest), using thumb and forefinger. If in doubt, leave well alone!

● *Where can I obtain more detailed information on local wildlife?*

Nowadays, there are faunal guides to virtually every major region of the world, as well as to local fauna and flora. General libraries are good places to start a search for relevant literature. However, for more practical guidance of a local nature, the best source of information is the natural history society for the area. Details may be found in the telephone directory, local pet shops, schools, libraries and citizen advice bureaux. As an additional and increasingly significant source of information, the relevant internet user groups that now operate on the Worldwide Web should not be overlooked, especially where urgent practical advice and support are required. These user groups, as well as several companies (including manufacturers and suppliers), often have helplines that are free of charge.

the shallows and snatch a fish, but this is an extremely rare occurrence. Raccoons, on the other hand, are nocturnal prowlers which will take fish from ponds on a regular basis. They wade into the shallows, and have a predilection for large, plump fish.

Controlling Mammals

Some mammals, such as foxes, are easier to control than others, like cats.

As the territory of foxes has extended into urban areas, so they have begun to lose some of the fear of humans they once had. Some urban foxes may even be so bold as to visit ponds during the day and may need to be chased off. Rural

foxes, however, remain timid and tend to retain their more nocturnal habits.

Again, as with insects, there are many people in the water gardening community who would argue in favour of adopting a *laissez-faire* attitude towards foxes. However, if – for any reason – control becomes absolutely necessary, then you should employ the same kind of deterrent as for herons and other birds, and reinforce this by chasing stubborn individuals away. Raccoons, too, can usually be controlled in the same way. In the case of persistent "offenders", it may be necessary to contact an animal control office or the relevant local authorities, who will trap the interlopers and relocate them.

Visiting cats are quite a different matter; as pets, they are liable to be around on a permanent basis and cannot be relocated. Controlling them is therefore more a question of cutting down the regularity of their visits than stopping them altogether.

Anti-bird fencing can be used to control cats, though the electrified wire must be set at a lower height (around 6in/15cm) than for herons. You can also try restricting easy access to the pond by planting a dense border of surrounding foliage, or by ensuring that the height between the potential vantage points and the water surface is more than 12in (30cm).

Of all the cat-control methods, the simplest is arguably also the most effective, though it does demand considerable commitment on the part of the pond keeper. It involves scaring cats off by shooing them away, and must be done on every single occasion that one is seen in the vicinity of the pond. Cats learn fairly quickly that they are not welcome inside your territory and will therefore tend to avoid it, or cross it swiftly and warily. You are still likely to lose the occasional fish, but this will become a rare rather than a frequent occurrence.

Where dogs are concerned, you should train your own pet not to enter the pond or to touch any fish or amphibians. It takes time and effort to achieve this, but patience usually brings excellent results.

In addition to the methods outlined above, cat and dog repellent sprays may be employed (these may also work against other mammals). Their effectiveness varies, but they are worth considering as an option. Do first ensure that the chemical compounds they contain are harmless to other animals, including amphibians and the creatures that form their natural diet, e.g. earthworms, beetles and other insects.

In the end, you may simply have to accept that excluding all wildlife from a pond is a practical impossibility. Control is the only realistic course of action. Besides, since being a successful pond keeper necessarily entails being a nature lover, it is probably more constructive to look upon the occasional visit by wildlife to your pond in a positive light, rather than as something to be prevented at all costs.

WILDLIFE AND THE LAW

Wildlife legislation exists to protect both the animals themselves and the humans who buy, sell or keep them. Wildlife protection has two main components. On the one hand, there are those regulations that control the actual collection of species deemed to be under threat and, on the other, those designed to control the introduction of species into habitats where they are known (or believed) to pose a threat to the native fauna and/or flora. Humans, for their part, are protected by various pieces of legislation that control or ban the keeping of certain dangerous animals.

The subject of wildlife legislation is a vast and complex one, with regulations that range in force from local to global levels, and which can change in as little as a few weeks, or months, or over several years.

International Laws The most relevant international trading laws concerning wildlife were first agreed in 1975, with the setting up of CITES (Convention on International Trade in Endangered Species of Fauna and Flora). Under the rules of the Convention, species are listed under three Appendices. Those that appear under Appendix I cannot be commercially traded, except under special dispensation granted by the CITES authorities if the specimens have been bred in captivity and the breeding programme meets certain stringent criteria.

The Coahuilan Box Turtle (*Terrapene coahuila*), for example, is an Appendix I reptile that cannot be traded in, and for which no captive-bred quotas are currently available, while the Dragon Fish (*Scleropages formosus*) can be exported under strict controls from a number of breeding establishments. These have been able to satisfy the CITES authorities that the specimens are being successfully bred and monitored over several generations.

Unlike *T. coahuila*, some other species in the genus *Terrapene* are listed under Appendix II, where regulated (licensed) trade is permissible. Appendix II contains a wider range of species than Appendix I and includes a significant number of reptiles and amphibians, from crocodilians, to some frogs, toads, newts and salamanders.

The most extensive Appendix in terms of species is Appendix III, where controls are not quite as stringent, but may still include certain international trade restrictions.

National Laws While CITES regulations aim primarily to protect wild species from over-exploitation, national authorities can take additional steps to protect their indigenous fauna and flora, by restricting or prohibiting both the collection, and the import of species or varieties regarded as a threat to native species.

National governments (in this case, Europe functions as a single nation) can even restrict or prohibit the import of species not listed under either Appendix I or II. The Red-eared Terrapin/Slider (*Trachemys scripta elegans*), for instance, is abundant in its native US range and is commercially bred in large numbers in some states (particularly Louisiana). However, for many years, its importation into Europe has been restricted or prohibited at certain times. The same applies to the American Bullfrog (*Rana catesbeiana*).

Local/State/Regional Laws Somewhat more restricted than national or international laws in their application – but, nevertheless, just as important – are local, state or regional regulations, which aim to protect fauna and flora over a more modest-sized area, or restrict or ban imports into, or exports from, such zones.

These laws can range from restriction or prohibition on ownership of certain species or varieties (e.g. Koi in some areas of Australia) which are seen as potential threats to local species, to the

actual movement of specimens across state boundaries, or the exportation of specimens below a certain size, e.g. small Red-eared Terrapins between US states, or the requirement that such specimens must be certified free of *Salmonella* bacteria.

Laws on Dangerous Animals There can also be regulations controlling or prohibiting the buying, selling or keeping of dangerous animals. Defining what constitutes a dangerous animal can be difficult, but, generally speaking, if a species is venomous or can cause injury, e.g. by biting, it is considered "dangerous".

This type of legislation is designed to protect, not just potential owners, but also other people who might be exposed to risk should such an animal escape. Where the buying, selling or keeping of a potentially dangerous animal is allowed, special licenses are likely to be required. Premises may also need to be inspected for suitability.

If you are a home owner, it is worth checking your conditions of property ownership for any clauses that might apply to the use of the premises, both indoors and outside, for accommodating dangerous animals. If you are a landlord and have such conditions written into your property deeds, these should be mentioned to potential tenants. Equally, if you are a tenant, check that – assuming you possess the appropriate licences – there are no other legal restrictions to your keeping such animals.

◄ *Bullfrog* (Rana catesbeiana).

Pond Management

ONCE YOU HAVE SET UP, STOCKED AND PLANTED
your pond or water garden, careful attention is
needed on your part if it is to remain in good condition
throughout the year. Smaller systems, especially when
heavily stocked with fish and plants, generally require a
more intensive management regime than larger, more lightly
stocked schemes. But the common factor of all ponds, however
large or small, is that they all need some kind of maintenance.

Prevention is certainly much better than cure when it comes
to pond and water garden maintenance. Although detecting and
correcting a mistake is undoubtedly satisfying in itself, it is always
far more rewarding if problems that are due to neglect or ignorance
simply do not arise in the first place.

To this end, the following pages offer practical guidelines to help
you keep your pond in the best of health. Since efficient water
management lies at the heart of all successful pond keeping, this
subject is treated first. The problems associated with particular times
of year are then highlighted, and the steps that you should take to
avert them are clearly laid out. Checklists summarize the key
points to watch out for in each season.

Naturally, no advice could hope to account for all eventualities,
and this book is no exception. What it does do is draw your
attention to the hazards that are most commonly encountered
by pond keepers and equips you to overcome them
well before they turn into a crisis.

▶ *On a frozen pond, Milfoil* (Myriophyllum *sp.*) *and
Duckweed* (Lemna *sp.*) *leaves lie covered in hoar-frost.*

Managing Water Quality

THERE IS A MAXIM IN POND KEEPING, TO THE effect that if we look after the water, then the water will look after our fish. There is a great deal of truth in this, since, without good water maintenance, it is quite impossible for fish and other pond occupants to enjoy long-term health.

Despite its simple chemical composition – two atoms of hydrogen to every atom of oxygen, or H_2O – water is a remarkable substance. It will dissolve or erode virtually anything, given enough time. The result of this inexorable process, plus the other compounds that are added through the metabolic processes of animals and plants, is that in Nature water never occurs in its pure H_2O state, but rather as a "soup" consisting of water molecules and a host of other dissolved and suspended compounds. Together, they impart a range of qualities to the water.

Of these substances, the ones that most concern the pond owner are those that make the water acid, neutral, or alkaline. Of especial interest are the potential killers of pond life, namely, compounds that contain dissolved toxins, such as ammonia and nitrites, and additives such as chlorine and chloramines, which make water safe for human consumption. In addition, the oxygen content of the water is also important. Water hardness – an essential consideration in aquaria – is also a significant factor in ponds, particularly where soft-water fish species are being kept (as might be the case in some tropical set-ups), or where hard-water fish species are kept in soft-water regions.

Acidity/Alkalinity

The acidic or alkaline properties of a solution (including water) are measured on a scale that ranges from 0 to 14, with the lower reading indicating extreme acidity and the upper one extreme alkalinity. Both are lethal to all forms of life. A reading of 7 represents water that is neither acid nor alkaline, i.e. it is neutral. This value is indicated as "pH7".

The scale is not linear, but logarithmic, which means that a change of one unit represents a tenfold increase or decrease. Therefore, a water sample with a reading of pH6 is ten times more acid than one with a value of pH7 and 100 times more acid than one with an alkaline reading of pH8.

Pond water ranges in pH roughly between 7 and 8, depending on the original source, the stocking level, the effectiveness of the filtration system, and other such factors. Fluctuations between these figures and values slightly above and below them are generally acceptable, as long as

they occur gradually. Even larger 24-hour fluctuations, such as those occurring in ponds that are heavily planted with oxygenators, can be tolerated by most fish, since the changes take place smoothly over this period, although controlling these swings is advisable.

The pH value of the pond water can be easily checked using a test kit. Several types are available, from those that use a colour chart to those that work electronically, giving a direct read-out. Therefore, if you have any trouble telling the difference between closely matched colours, you should opt for an electronic meter.

Once a pond has had a chance to settle down, it should only be necessary to test the pH about once a week, until a series of steady readings has been obtained. After that, the tests can become less frequent.

● *What is the best way to balance out fluctuating water pH levels?*

... Ponds that are heavily stocked with submerged oxygenating plants can experience fluctuations in acidity or alkalinity over a 24-hour period. In extreme cases, fish may suffer a shortage of dissolved oxygen during the night, especially during warm, humid weather, and appear gasping at the surface. (see Why are Plants Important?, pages 118–119). To balance fluctuating pH levels, first take an oxygen reading to make sure that the gasping is caused by low dissolved oxygen levels. If so, conduct a partial water change, and increase the aeration level, by adjusting the flow from your fountain or other water feature, and by turning the venturi (aeration) attachment on your filter return to maximum. Further remedies are to play a fine spray from a hose onto the water, and to thin out the oxygenators. Monitor oxygen levels until they stabilize at at least 6mg per litre.

▲ *Electronic meters make pH testing both easy and quick, and are especially useful for people who have difficulty distinguishing between the colours used in standard pH charts.*

◀ *A well-balanced pond such as this will not experience major day/night pH fluctuations.*

Oxygen

Pond fish require around 6mg/l of dissolved oxygen for stress-free respiration. In an adequately stocked pond which has an efficient filtration/aeration system, shortages are unlikely to occur.

However, as the temperature of water rises, its oxygen-holding capacity decreases. Consequently, warm, still summer nights can cause reduced oxygen levels, especially in ponds that are heavily stocked with both fish and plants. A sudden drop in barometric pressure, such as happens before a thunderstorm, can have a similar effect.

Oxygen testing kits and electronic meters are widely available; the latter are a little easier to use. Readings should be taken regularly, especially during summer, if an oxygen shortage is suspected. The best time to take a reading is at daybreak, when the oxygen level is at its lowest.

Ammonia/Nitrites/Nitrates

Of these three compounds, the first two are especially toxic and can kill, even at relatively modest concentrations. Ammonia, for example, can be toxic to some delicate fish species at levels as low as 0.025mg/l, and can kill relatively quickly at concentrations of 0.2–0.5mg/l, particularly under high pH conditions. Nitrites are less toxic, but can still cause damage at 0.1mg/l. Nitrates, however, can rise to around 50mg/l before even delicate species begin to show signs of distress.

The presence of these substances can be measured with test kits or electronic meters. Monitor levels at regular intervals, especially during the early days of maturation, when the bacterial population of the filter is being built up in tune with the gradual stocking of the pond with fish.

Water Hardness

Mg/litre CaCO₃ (parts per million)	Description
0–50	Very soft
50–100	Moderately soft
100–150	Slightly hard
150–200	Moderately hard
200–300	Hard
Over 300	Very hard

Hardness

Salts (particularly calcium and magnesium) dissolve in water and make it either soft or hard.

Although soft-water species of fish can adapt to hard water and vice versa, each should, ideally, be kept in water suited to its natural requirements. Most pond fish live perfectly happily in medium-hard or even slightly harder water, as this puts less stress on the metabolic processes that control the water balance of the body tissues (osmoregulation). In soft-water conditions, these fish will adapt accordingly, but may be further helped by the addition of pond salt to the water (follow the manufacturer's recommended dosage).

Testing kits and electronic meters to measure water hardness are widely available and should be used to ascertain the natural hardness of the pond water, both during the initial stages of setting up the pond and subsequently on an occasional basis, thus allowing you to decide whether or not it should be adjusted to meet the general requirements of proposed inhabitants.

Tropical species have varying water hardness preferences. These should be checked in an appropriate aquarium reference book prior to their introduction to the pond.

Chlorine and Chloramine

These substances are added to tapwater to render it safe for human consumption. However, both are toxic to fish.

Chlorine is quite volatile and will disappear from the water, particularly under vigorous aeration, in a day or so. Chloramines are more stable and can take 7–10 days to break down, if this is left to natural processes.

Dechlorinators and dechloraminators should be added to the water whenever a partial water change is conducted, particularly if this amounts to around 15–20% or more of the total volume of the pond. However, there is little point using a dechloraminator if your local water supplier does not use chloramines, so find out in advance whether this is the case.

Even if chloramines are not used during the winter months, they may be added to water supplies during hot weather so, again, check frequently during the summer season.

▲ *Testing ammonia levels in a pond. Concentrations of this compound will fluctuate during the maturation period, but should stabilize thereafter.*

● *Will "acid rain" or "fallout" from nearby factories affect my fish?*

... Ponds located close to some types of factory may experience slightly higher (but not dangerous) levels of some chemical compounds in the water. If pH testing shows this to be the case, it would be best to avoid delicate species such as Orfe (*Leuciscus idus*), which may experience some distress when such conditions occur in conjunction with low oxygen levels.

● *Recently, I have seen the fish in my pond jumping above the water surface. Can you suggest a possible explanation for this?*

Although fish will occasionally jump when being chased, if they do so repeatedly, the more likely explanation is that there is something wrong with the quality of the water. Increased levels of ammonia or nitrites are frequently to blame; after testing for these, carry out a partial water change, and check that the filter is working efficiently. If you have recently added new fish, stop feeding and allow the filter to process the additional waste. Then continue to monitor ammonia and nitrite levels on a daily basis. If no improvement is evident, either reduce fish stocks to their former level, or upgrade your filtration system. Only resume feeding once toxin levels have dropped.

Routine Maintenance

Keep a close watch on water quality at all times, irrespective of the time of year, as this is central to the survival of fish and plants.

It is difficult to stipulate how often pH, ammonia, nitrite, nitrate, oxygen and hardness tests should be carried out, since a great deal depends on temperature and other climatic conditions, stability in the quality of the supply, stocking levels, frequency and types of feed, as well as many other factors. However, when a pond is first set up, pH, ammonia and nitrites can be tested for daily during the first couple of weeks to see how they are changing and to give you an idea of how well your pond can cope.

As the filter matures and the plant population begins to grow, weekly tests will suffice. Thereafter, once a safe balance has been established, monitoring should be regular, but less frequent.

Adopt the same approach to oxygen levels, adjusting fountains and filter returns, or waterfall, stream or cascade flows until a satisfactory level has been achieved. At other times, testing may only be necessary, or desirable, if the fish show signs of distress or during periods of prolonged low barometric pressure.

Hardness may be measured when the pond is first set up and occasionally after that. Alternatively, details may be obtained from your local water supplier, who will also be able to provide information on chlorine and chloramine levels.

✓ TIPS ON WATER QUALITY

- ✓ Do not overstock with either fish or plants.
- ✓ In the early stages after maturation, stock the pond with only half (or less) of its eventual stocking quota. Increase this gradually over a period of at least several weeks.
- ✓ Check water parameters regularly (more often in the early stages) and adjust.
- ✓ Ensure efficient pump and filter operation.
- ✓ Follow a year-round maintenance routine.
- ✓ Resist the temptation to overfeed your fish.
- ✓ Make sure that all fountains, venturis and other aerating devices are running properly, especially in heavily planted ponds and during periods of low barometric pressure.

Winter Maintenance

THE MANAGEMENT OF A POND DURING WINTER depends to a large extent on its location. In the tropics, the difference (in temperature, though not necessarily in rainfall) between winter and summer is small compared to the climatic fluctuations that are experienced in temperate or high-altitude zones.

In tropical climes, therefore, winter maintenance is more a question of reducing activities like feeding – but only if the fish indicate that this is necessary by displaying a lack of appetite. Thinning out of containerized plants, overhauling the pond equipment and controlling leaf-fall may also be carried out at this time.

In more temperate regions, seasonal climatic characteristics dictate that maintenance routines must be altered to cope with the pronounced changes that take place during the colder months of the year. The two most noticeable changes are shortened (and often darker) days and lower temperatures, particularly at night.

When these factors begin to occur on a daily basis, rather than just occasionally, fish, other creatures and plants respond in various ways. Fish will move into the deeper areas of the pond for longer periods, while plants – most of which will have stopped growing and begun dying back during autumn – will die back to their maximum extent and enter their resting phase.

Wise Precautions in Winter

As winter progresses, ice may form on the water surface with increasing frequency. In the coldest spells, the ice cover may be several centimetres thick and may not thaw out at all during daylight hours. Snow, too, may cover the pond, contributing further to the already dark conditions within the pond itself, by impeding light penetration. If a thick, persistent layer of snow covers the pond, sweep this away (but without stepping on the ice!) to allow the light to penetrate.

As temperatures drop, the fish in the pond will progressively lose their appetite and will cease feeding altogether once the temperature reaches around 4–6°C (39–43°F). No attempt should be made to feed them at such times. Indeed, in temperate zones with freezing conditions, many pond owners do not feed their fish at all during the winter months. In others, the feeding programme should be light and consist of easily digestible formulations.

◄ *A frozen, snow-covered pond in the depths of winter. Careful management is required during this potentially dangerous time of year if your livestock is to survive in good health.*

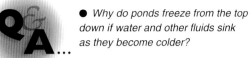

✓ Always ensure that the pump is placed near to the filter to avoid undesirable currents.

✓ Raise the pump off the bottom of the pond, seating it securely on bricks or breezeblocks – this allows some of the warmest (relatively speaking) water to remain where it is most needed. It also prevents debris from being stirred up by the pump and clouding the water.

✓ Locate the filter return below water level – warm return water splashing on the pond surface will cool it down.

In order to allow toxic carbon dioxide (which can kill fish) to escape from under the ice, a hole must be kept open on the water surface. The easiest method of doing this is by using a floating pond heater and keeping the pond pump running. However, other alternatives include: floating a squeezeable ball on the surface (removing it each morning); floating weighted polystyrene boxes or plastic bottles filled with warm water that can be thawed out; or simply pouring hot water from a kettle directly onto the ice. On no account should a hole be smashed through the ice. The shock waves can stun the fish or even kill them in extreme cases.

Some pond owners switch their pond pumps and filters off during the winter in cold regions. If you do this, both the pump and the filter should be overhauled and the filter allowed to mature again once it is re-activated in the spring.

Whole-pond heating – as opposed to the use of a floating pond heater to maintain an ice-free hole – is becoming progressively more popular in temperate zones, particularly among Koi keepers. The aim of such heating is to maintain the temperature of the pond water above 10°C (50°F) throughout the winter, thus keeping the surface of the pond ice-free and the fish active and feeding throughout this period. Fish that are kept under these conditions will

▶ *Pond heaters provide an easy way of maintaining an ice-free hole, but are not powerful enough to raise the water temperature throughout the pond.*

Q&A...

● *Why do ponds freeze from the top down if water and other fluids sink as they become colder?*

Although cold water sinks, there is a "cut-off" point that comes into play when the temperature drops to 4°C (39°F). When this happens, individual water molecules begin to become more "organized" and their structure more open. The water thus becomes less dense and floats. Finally, as the temperature continues to fall towards 0°C (32°F), the openness and organization of the molecules increases further, and the whole layer of water molecules spontaneously freezes.

● *Is it a good idea to overhaul a pond in winter?*

From a fish's point of view, winter is just about the worst time of year for an overhaul. As "poikilothermic" creatures, their body temperature drops gradually in winter in tune with that of the surrounding water. The last thing that fish need is to be chased, netted, put in temporary quarters, or otherwise distressed. As a further source of stress, a few days later – when they have only just managed to adapt to their new surroundings – they get poured back into a pond containing water of a totally different chemical composition, not only to the water in the pond before the clean-out began, but also to that in its temporary accommodation. Not surprisingly, some fish don't survive this traumatic experience and fall victim to "Spring Sickness" later on. This is partly caused by stress and partly by being forced back into hibernation, which gives them no opportunity to rebuild their depleted energy reserves.

● *Are pond liners damaged by frost?*

Only the cheaper, less flexible and less durable types made of polythene are likely to suffer this form of damage. Good-quality butyl rubber liners (or their equivalent) that carry a guarantee are not susceptible to frost damage. They remain pliable – albeit with a little stiffness – throughout even hard winters in temperate regions.

▲ *Appearances can be deceptive – despite the freezing temperatures experienced in winter, life still goes on underneath the ice, but at a much slower rate than at other times of the year.*

continue to grow and will not be exposed to the stress that is caused by going into hibernation and re-awakening in the spring after a lengthy period of torpor, during which energy reserves are gradually expended.

Whole-pond heating also means that pond keepers can maintain an active interest in their hobby throughout the winter. Moreover, the fish do not appear to suffer as a result of missing out on the period of hibernation that they would normally undergo, either in the wild or in non-heated ponds. On the other hand, heating costs are not inconsiderable and, since adequate pre-winter conditioning of fish allows them to over-winter safely anyway, whole-pond heating may be seen more as a matter of personal choice than biological necessity.

Note that 10°C (50°F) is the minimum temperature that whole-pond heating must maintain, even during severe weather conditions. At lower temperatures than this, the fish will still be active and will feed. However, their immune system will only be ticking over and so will not be strong enough to fight off bacterial and/or viral infections effectively. Though these microscopic pathogens function at a slower rate in colder temperatures than they would in warmer water, their metabolism is still working and so they reproduce. As a consequence, they represent a health threat to your fish, even at these relatively low temperatures.

✓ YOUR POND IN WINTER

- ✓ Stop feeding the fish when temperatures drop to around 4–6°C (39–43°F).

- ✓ Create some shelter for the fish, e.g. with a stiff plastic sheet held between bricks or a dark plastic bucket resting on its side.

- ✓ Maintain an ice-free hole at all times.

- ✓ Bring very small fish from late hatchings indoors, if you haven't already done so during autumn.

- ✓ If a large, external biological filter is kept running throughout winter, a small aquarium heater or thermostat unit installed in the settlement chamber will help keep bacterial activity going during the coldest weather.

- ✓ If the pump and filter are being decommissioned for winter, service the pump and clean the filter. Use pond water on the filter; detergents will kill beneficial filter bacteria.

- ✓ If the pond is used solely for fish, (e.g. many Koi set-ups), don't switch off the pump and filter system, but run it normally or at a reduced rate (if the latter, check toxin levels).

- ✓ Overhaul all remaining pond equipment that is not in use.

- ✓ Protect frost-sensitive plants with mulching or sacking.

- ✓ Bring the more tender plants indoors (many marginals make good house plants).

Spring Maintenance

IN TEMPERATE REGIONS, THERE IS NO MAGICAL sudden re-awakening as winter turns to spring. Instead, it is a gradual process, but one that shows distinct changes from one week to the next. Activity increases on all fronts, from the arrival of frogs and toads in preparation for breeding, to an increase in energy among the pond fish and the sprouting of shoots on the early-growing plants.

All these signs of renewed life are, however, accompanied by risks, since this time of year is also characterized by wide temperature and climatic fluctuations. Warm, sunny days are often followed by cold, chilly nights that can damage tender new growths and have a detrimental effect on the health of fish.

The immune system of fish will gradually start working once the water temperature reaches around 10°C (50°F). It is therefore vital not to create situations that will place the fish under undue stress, thus laying them open to attack from pathogens. Provided that both your fish and pond were adequately prepared for winter during the previous autumn and that the basic principles of winter maintenance have been adhered to, the risk of "Spring Sickness" (see page 191) should not be too high. Even so, you should keep a close watch and take appropriate action if necessary (see also Health Management, pages 74–81 for details of specific diseases).

Feed your fish sparingly at first, providing them with an easily digestible diet given early in the day to allow the fish to process the food before temperatures fall in the evening. A heavier formulation and more frequent feeds may be introduced once water temperatures begin to settle above the 10–12°C (50–54°F) mark for most of the day.

▶ *Early spring is a good time to give pond equipment such as pumps and pond filters an overhaul in readiness for the coming season, especially if they have been kept running throughout the winter.*

Gradually turn up the water flow through the pump and filter and carry out regular testing for ammonia and nitrites, reducing the amount of food given to the fish if either of these compounds show a marked increase. The ideal situation to aim for is an increase in food without an increase in toxins. The use of a bacterial filter feed, or the temporary incorporation of a zeolite (ammonia-adsorbing) package within the filter system, may help at this time.

If either ammonia or nitrites show an increase (whether or not the above procedures are followed), carry out a partial water change, adding a dechlorinator or dechloraminator to the fresh water if the change involves more than around 10–15% of the total pond volume. In any case, a spring flush-through of the pond will help clear out any undesirable compounds that may have accumulated during winter.

▲ *Removing blanket weed (filamentous algae) from the pond is an essential spring task. You can also keep this pervasive nuisance down with an eco-friendly algicide.*

✔ YOUR POND IN SPRING

- ✔ Increase feeding gradually, using light formulations at first.

- ✔ Monitor health status of fish regularly and take appropriate action.

- ✔ Provide temporary shelters for fish.

- ✔ Test water on a regular basis for ammonia and nitrites.

- ✔ Increase flow rates through pumps and filters.

- ✔ Flush out the pond or conduct a partial water change (use dechlorinator/dechloraminator).

- ✔ Redesign planting arrangements and upgrade pond systems.

- ✔ "Seed" filters with bacterial feed if desired and/or add a zeolite (ammonia-adsorbing) component to the filter system.

- ✔ Divide or propagate plants as necessary or desired (see Pond Plants, pages 120–123 for further details).

- ✔ Control algae by use of ultra-violet sterilizers, algicides, and/or physical removal.

● *Is spring a good time to clean out a pond?*

Q&A ... Yes, there are several advantages in carrying out a spring pond clean-out. Many plants will be able to tolerate moving at this time of year, splitting and replanting much more comfortably before they really begin to grow in earnest. Fish will have come out of hibernation and, once feeding and fully active, will be able to handle the clean-out much more easily than during winter. However, complications can arise if spawning activity is under way, or shows signs of getting under way. On the one hand, if the clean-out is delayed until spawning is over, the season can have progressed to the point where the overhaul is no longer feasible. On the other, if the clean-out goes ahead despite spawning activity, it is then only fair to provide the fish with spawning facilities such as spawning brushes, mats or clumps of fine-leaved vegetation, in their temporary quarters.

● *How often does a pond require a spring clean-out?*

If properly maintained, plantless ponds, such as Koi pools, can go for many years without requiring a complete clean-out. General or mixed ponds that are adequately stocked with fish and plants and properly maintained will usually require an overhaul every few seasons. It is difficult, though, to put an exact timescale on the interval between clean-outs. The best guides are the pond itself, its inhabitants and the water. If a pond does not "brighten up" after a spring partial water change, and the water looks dull and dark instead of crystal clear, it may be time for an overhaul. Other signs that a clean-out is due are if the water gives off an unpleasant smell, if large amounts of sediment are churned up by the fish, or if bubbles float up to the surface on a regular basis from the bottom mulm.

● *What is the best way of re-activating a filter that has been switched off during winter?*

If a filter has been left running, even at a low level, during winter, its bacterial population will often be able to cope with the increased biological load in spring. Even if a filter has been switched off and cleaned, there will still be a coating of bacteria. However, this population will be much lower than in a filter that has been operating throughout winter and will therefore greatly benefit from the addition of a proprietary bacterial-aid preparation. As a further help, increase gradually the amount of dissolved wastes that the filter has to detoxify through controlled feeding of your fish. A packet of zeolite placed within the filter system will help adsorb excessive ammonia until the filter bacteria are sufficiently well established to cope on their own.

As sunlight intensity increases, there may be an algal bloom, resulting in "green water" (free-floating algae) and/or clumps of blanket weed (filamentous algae), often following a partial water change. Resist the temptation to clear this by carrying out an additional water change, as it will only provide the algae with a fresh supply of nutrients. Free-floating algal blooms can be kept under control with an ultra-violet sterilizer, but filamentous algae will require regular removal. Once pond plants becomes established, the risks of algal infestation will decrease or disappear. A pond algicide may also be used in conjunction with other treatments, but check the manufacturer's instructions and follow these precisely.

Spring affords an opportunity to redesign the planting arrangement and install extensions to the existing system. Yet it is also a period – particularly during the early weeks – when the lack of vegetation may expose the fish to attack from predators. It is therefore a good idea to provide some artificial shelters – for example flower pots, pipes or ledges – where they can hide.

Spring Sickness

"Spring Sickness" is not, in itself, a distinct fish disease. It is a term used to refer to a general state of poor health exhibited by some fish during spring. The actual disease may be anything from bacteraemia (mild, internal, bacterial infection) to excessive body slime production caused by an increased skin parasite load, or any combination of factors that produces symptoms of ill health at this time of year. Many of the lighter forms of Spring Sickness, like excessive body mucus production, will clear up of their own accord, once the immune systems of the affected fish have built up some momentum. At least, this is the case with fish that have overwintered successfully and are in a good state of health. Fish that have come through winter in a weakened state will experience greater difficulty, and may therefore either succumb to the effects of the original infection, or become susceptible to others. It is these fish that account for the large majority of deaths that are generally ascribed to Spring Sickness. In order to minimize the health risks associated with spring, many pond keepers treat their ponds with a general anti-parasitic and/or anti-bacterial compound. However, if the latter is being used, the dose must be light to avoid damaging beneficial filter bacteria in the process. If bacterial problems are suspected, it is best to seek advice from a vet, who will be able to prescribe a treatment of flakes or pellets that have been impregnated with antibiotics.

▶ *"Green water" results from high levels of illumination and dissolved nutrients such as nitrates, and blights most ponds at some stage with its "pea-soup" appearance.*

▼ *Excessive algal growth requires careful handling. Dosing heavily with algicides, but without remedying the underlying cause, will only give temporary respite.*

Summer Maintenance

SUMMER IS THE SEASON WHEN THE MOST sustained activity takes place in and around ponds, after the relatively short, but intense, spring surge. Plant growth is at its most vigorous and feeding demand is at its peak. Feed several times per day, but follow the "Five-Minute Rule" on every occasion (see Q&A, page 196).

As a direct result of this increased activity, filters have to be able to cope with the added waste load. Testing for ammonia and nitrites will show whether or not your filter is managing to keep up. If it is not, you may need to reduce fish stocks, scale down the feeding programme, or increase and improve the filter capacity. Regular cleaning of pre-filter sponges or the equivalent should also be carried out during this season. Prolonged high summer temperatures can cause fish to lose their appetite, as a result of excessive metabolic demands in an environment where oxygen levels may be lower than normal (see below). If fish

▶ *Regular summer maintenance includes curbing plant growth by cutting back or thinning out, to prevent them from blocking water courses and polluting the water.*

exhibit a lack of interest in food at such times, resist the temptation either to medicate or to offer alternative food. You may find that their appetite is at its peak during the early morning and late evening, so feed at these times if the above "symptoms" become apparent.

Plant pest activity is at its most vigorous during summer. Keep a daily watch, and take action promptly, since, under warm summer conditions, problems can spread at an alarming rate. As a first step, remove all dead and damaged leaves and blooms on a regular (preventive) basis; also check in Plant Health and Diseases, pages 128–129 for details on how to deal with particular pests and diseases. Even in healthy plants, dead leaves and blooms should be removed regularly, since otherwise they will rot in the water, further adding to the biological load on the filtration system. Healthy plants may also be thinned out as appropriate or as desired. Water-borne and other predators may also be at their most active at this time of year. Monitor the situation; one sign of a visit by a predator is fish hiding at the bottom of the pond.

Evaporation from ponds is a familiar summer problem, especially if there are streams, fountains, cascades and waterfalls and if the pond has little surface cover in the form of water lilies and other similar plants. Water is lost from the pond in two ways. Firstly, marginal plants in particular take up large quantities of water through their roots during the growing season, and lose it through their leaves (a process known as "transpiration"). Secondly, high air temperatures will cause water loss through direct evaporation from the pond surface. Regular top-ups will be necessary. If these are small and frequent, the water will not generally require dechlorinator/dechloraminator treatment.

✔ YOUR POND IN SUMMER

- ✔ Feed fish often on a high-nutrition diet.
- ✔ Prepare indoor accommodation for at least some of the fry that are produced as a result of late spawnings.
- ✔ Maintain all pond filters at maximum operating capacity.
- ✔ Test water regularly and carry out necessary modifications.
- ✔ Continue with the programme of planting/thinning out/propagating.
- ✔ Monitor plant health.
- ✔ Top up the pond water to replace water lost through direct evaporation from the pond surface and transpiration via plants.
- ✔ Remove dead leaves/blooms regularly.
- ✔ Check oxygen levels during humid nights and increase aeration if necessary.
- ✔ Watch for predators and take appropriate action (see Other Wildlife, pages 172–179).

Summer Filter Operation

✓ Keep solid wastes away from the biological areas of the filter by using settlement chambers and by regularly vacuuming the pond bottom.

✓ Use a venturi attachment (aerating device) in the clean-water return from your filter to raise oxygen levels in the pond and promote aerobic bacteria activity in the filter.

✓ Regularly rinse biological filter media to prevent clogging.

✓ Stir up gravel and granular-type filter media to prevent compaction and "canalization", i.e. the formation by accumulated debris of narrow, free-flowing channels that impair efficient operation of the filter.

Q&A

● *What is stratification?*

A ... As the temperature rises during the day, the topmost layers of water are the first to warm up. Where a stream, fountain or other form of moving water is incorporated within a scheme, this warmth is evenly distributed throughout the pond. Yet, if the pond has no water movement, or very little, the warm layer will tend to rest on top of the lower, cooler and deeper water. The resulting layering effect is known as "stratification". In most ponds, this won't create a major problem, but in deep designs housing round-bodied types of fancy Goldfish, the rapid chilling that can occur as the fish repeatedly dive to the bottom may change the gas balance within the swimbladder which, in turn, will disrupt buoyancy control. Fish affected in this way will sink, float or exhibit other forms of swimming imbalance.

▶ OVERLEAF *A well-kept garden pond can look resplendent at the height of summer, with its plants displaying lush foliage and colourful blooms.*

▲ *Clearing Duckweed from a large ornamental pond. This vigorous floating weed is at its most invasive during summer, and must be removed regularly.*

Dissolved oxygen levels may drop during hot, humid nights, and aeration, via a fountain or venturi attachment, should therefore be left running at all times. If these attachments are not available, directing a fine spray of water over the pond from a garden hose will help. This is particularly effective in late evening, as by this time the oxygen production that takes place during daylight hours (as a result of photosynthesis by submerged plants) has ceased.

Many fish species will continue to spawn well into the summer. If spawnings occur during late summer, it may be necessary to prepare alternative indoor accommodation for at least some of the fry to enhance their chances of survival over the coming autumn and winter. This mostly applies to areas where cold-season temperatures are likely to drop towards the lower tolerance level for the species in question. However, hatchlings from mid-season spawnings should be somewhat tougher by the time the colder weather begins to set in.

Q & A

● *How often should fish be fed during the summer?*

... During the summer, fish will require more frequent feeding, and in larger total daily amounts, than at other times of the year. During excessively hot weather, though, some temperate species like Koi and Goldfish can lose their appetite (see Foods and Feeding, pages 62–67 for further details). At their summer peak, pond fish can manage as many as five or six meals per day. However, the amount given at each feed should be such that it can be completely consumed in no longer than five minutes or so (the "Five-Minute Rule"). If it is not possible to feed fish this frequently, two to three feeds will be perfectly adequate, particularly in a pond that is well stocked with plants and their associated animal contingent. Some commercial fish breeders install automatic "demand feeders" for the duration of the peak feeding and growing season, and these may be worth considering if the pond is heavily stocked with fish that cannot normally be fed several times a day.

● *Can ponds overheat in summer?*

If a pond is at least 18in (45cm) deep, it is unlikely that it will ever become intolerably warm for its occupants. Very shallow designs, sited in full sun in hot regions (or during a long heatwave in temperate areas) may, however, experience such problems.

Autumn Maintenance

As LATE SUMMER BEGINS TO GIVE WAY TO early autumn, plants are usually the first elements of the pond to show signs that the most active season is drawing to a close. For instance, oxygenators may start to look straggly, while bog plants, marginals and surface plants will either stop growing entirely or show early signs of die-back. The fish, on the other hand, are likely to be almost as active as at the height of summer and will still be feeding avidly.

This is the time when early preparations for winter should be put into effect in temperate zones. Oxygenators should be thinned out or pruned back if their growth has been excessive. Shoots that have reached the surface should be trimmed back, since they may be killed off by ice later on. Dead leaves and blooms should be removed from other plants before they have a chance to detach themselves and sink to the bottom of the pond.

As the season progresses, but before any frosts set in, some bog plants, marginals and surface plants may be divided. Species and varieties that are tenderer, such as the Arum Lilies (*Zantedeschia* species and varieties) or the Umbrella Plant (*Cyperus alternifolius*) should be transferred to deeper water to protect them from ice, brought indoors, or sheltered in a greenhouse. Tender bog plants should be given a protective mulch of straw or equivalent. Tropical lilies should be lifted and stored as outlined in Surface Plants, page 152, or treated as annuals and replaced with new plants the following season. Tender floaters, for example the Water Hyacinth (*Eichhornia crassipes*), should also be brought indoors or treated as annuals.

Any leaves that drop onto the pond must be removed regularly (even several times a day) to stop them sinking and rotting at the bottom. Covering a pond with netting may look unsightly, but it is the most effective way of keeping leaves off the water, and should therefore be seriously considered as an option during the few weeks when leaf fall is a major problem.

As long as they still appear eager for food, fish should be fed several times a day, but late evening feeds should be discontinued, as autumn is characterized by unpredictable fluctuations in temperature. Their diet should gradually be changed from heavier summer formulations (usually protein-based) to a lighter cold-weather one (usually carbohydrate-based). Cease feeding altogether as soon as the fish appear reluctant to clear up the food within about five minutes.

Some fish may have spawned in late summer, especially if a partial

◀ *Late flowers, such as this Water Lily bloom, can be damaged by early frosts, and must be removed during seasonal maintenance.*

water change or a substantial top-up was carried out. The resulting fry may not be robust enough to withstand a harsh winter outdoors, so bring at least some of them indoors, housing them either in an aquarium, or in a vat or trough in a frost-free greenhouse.

Pumps and filters should be serviced before the really cold weather sets in. If at all possible, use pond water to rinse out the filter media, so that the beneficial bacterial population is not destroyed. If more than one biological medium is being used, rinse only one per week, to optimize continuity of the bacterial population. Similarly, if the filter has more than one chamber, clean one of these out each week. If temperatures fall markedly on a continuing basis, change over early to the winter pump and filter arrangement (see also Winter Maintenance, pages 186–188).

The pond can be given a partial water change or flush-through, but only during early autumn while the fish are completely active and the water temperature is still mild. A total pond clean-out can also be undertaken, but again this must be done early to avoid any problems associated with cold temperatures.

One of the principal objectives of the various tasks outlined above is to prevent the pond and its surroundings from deteriorating to the point where conditions will adversely affect the health of the fish and other pond occupants. Autumn maintenance also helps prepare the fish themselves for the rigours of the coming winter.

▲ *Netting hardly enhances the appearance of a pond, but it is only a temporary measure, and often represents the best method of controlling falling leaves. If left on the pond (inset), foliage can rot and affect water quality.*

✔ YOUR POND IN AUTUMN

- ✔ Trim back, prune or divide plants.
- ✔ Protect the more tender species and varieties of plant, or discard and replace them the following spring.
- ✔ Remove leaves that fall into the pond on a regular basis, or net the pond during peak leaf fall period.
- ✔ Collect and store winter buds (turions) as necessary (see Pond Plants, pages 116–161 for specific details).
- ✔ Slowly change fish food from summer formulations to lighter cold-weather formulations.

- ✔ Offer some protection to at least some of the young fish from late spawnings by bringing them indoors.
- ✔ Carry out a partial water change or a total clean-out of the pond before the cold weather sets in.
- ✔ If a pond is being constructed or installed during autumn, do so before frosts begin and the ground hardens.
- ✔ Service all pumps and filters.
- ✔ Move to the winter pump/filter arrangement earlier than scheduled if the weather deteriorates markedly.

In Nature, there are various reasons why falling leaves and accumulated debris do not normally have deleterious effects on the inhabitants of ponds. Firstly, pools in the wild usually contain relatively large volumes of water and small populations of fish. In addition, those fish that are present are tough, wild types that have evolved to cope with the prevailing environmental conditions. None of these important factors generally apply to ponds in gardens or on patios, hence the need for an adequate maintenance regime.

Apart from maintenance, early autumn is also a good time for pond installation, when the ground is moist but not waterlogged, and workable rather than frozen. One advantage in constructing and installing ponds in early autumn is that they then have several months to mature before they come to be fully stocked the following spring. However, there are also some drawbacks (see Q&A section right).

Q & A...

● *How often should leaves be removed from a pond during autumn?*

If leaf fall is considerable, you may need to skim them from the pond several times a day. However, much depends on the leaves themselves. Clearly, if they are of a type that becomes easily waterlogged, you will need to remove them more frequently than if they are water-resistant. If a pond is netted or otherwise covered, the interval between removals can be as long as a week, or even longer if the fall is not excessive. This can be prolonged further by stretching the net over the pond and letting the wind blow off at least some of the leaves.

● *What are the main advantages and disadvantages of constructing or installing a pond during the autumn?*

One distinct advantage is that the pond will have the benefit of several months of maturation before it is fully stocked the following spring. If it is constructed early enough, the pond can be stocked with a reasonable selection of plants and a few fish which, provided they are conditioned properly before you purchase them, will settle down and continue feeding until the cold weather sets in. The filter, too, will have a chance to develop at least some of its bacterial population, especially if it has been seeded with a proprietary bacterial start-up preparation. On the downside, plants introduced during autumn are at the end of their season, and so will not exhibit any signs of growth for several months. Planting therefore requires more imagination during autumn, since you will have to visualize the final effect. Stocking with fish must be very carefully controlled during autumn, as the colder temperatures slow down the maturation of the filter bacterial population, irrespective of any added assistance provided by start-up preparations. Therefore, if an imbalance should occur at this time of year, resulting in health problems for the fish, they will have little time to recover fully before the arrival of winter.

● *Is autumn a good time to install a pond heater?*

Certainly. If the heater is thermostatically controlled, it will come into operation once temperatures drop to a pre-set level (usually around freezing or just above). This means that as soon as ice begins to form on the water surface, the heating element switches on automatically and will maintain an ice-free hole around itself. Such an arrangement is preferable to waiting for ice to form and then having to create a hole to install the heater.

Glossary

Activated carbon A substance used in mechanical and chemical filtration systems to remove dissolved products (pollutants) from water.

Adsorption The process by which molecules adhere to the surface of a substrate, e.g. ammonia on activated charcoal or zeolite.

Aerobic Requiring oxygen.

Algae Primitive plants, either microscopic or large (e.g. kelp), which are almost exclusively aquatic and do not bear flowers.

Ammonia (NH_3) A gas that dissolves in water and is the first byproduct of decaying organic material; also excreted from the gills of fish. It is highly toxic to fish and invertebrates.

Anaerobic Not requiring oxygen.

Barbel Filamentous extensions, usually surrounding the mouth, by which the fish detect (taste) food.

Biorhythm Cyclical intrinsic behaviour or metabolic change in living organisms, e.g. seasonal breeding behaviour, day/night activities.

Black water Water that has a high level of humic acids and low levels of nutrients.

Brackish water Water that is intermediate between fresh and marine in its level of salinity.

Buffering action The means by which a liquid maintains its pH value.

Bulbil A small bulb-like structure, bud or swelling produced by some plants at the joint between a leaf or branch and the stem. Bulbils may fall off and produce new plants.

Canterbury spar Angular, porous gravel that can be used as a mechanical and biological filtration medium.

Carnivore A flesh-eating animal or plant.

Centrifugal force The force that causes the outward spinning movement of molecules (e.g. water) as the material is thrown outwards from the centre.

Centripetal force The force that causes a spiral movement of molecules that drags the whole material towards a vortex created at its centre, as when bathwater swirls down the drain.

Circadian A term applied to biorhythms that are approximately one day in duration.

Cone cells Cone-shaped cells found in the retina of animals. These cells are responsible for colour perception. Other cells, called rods, are responsible for perceiving shapes.

Consumption A general term used to refer to a condition that involves severe loss of weight. Fish suffering from consumption have thin bodies and overly large heads.

Crepuscular Active at dawn and dusk.

Cyanobacteria A primitive life form that has some characteristics of both bacteria and algae but is regarded as being different from both. Also commonly referred to as "blue-green algae" or "slime algae".

Dioecious A plants that has male and female flowers on separate individuals. In monoecious plants, both sexes occur on the same specimen.

Distichous A structure that is arranged in two rows.

Eft Newt tadpole stage.

Emersed Rising above the water surface, e.g. aerial shoots, leaves, and blooms.

Enzyme Protein-based biological catalyst that influences metabolic reactions.

Epidemic Outbreak of a disease that affects numerous individuals at the same time.

Exotic Strictly speaking, any species that is not native (indigenous) to the area in which it is present. In more general usage, it has come to refer to species that are rare or have highly distinctive characteristics.

Fastigiate Bearing erect branches close to the stem.

Flocculation The clumping of small particles into larger pieces.

Fry Young fish.

Gamete A sex cell, i.e. sperm or egg.

Gamma irradiation The technique by which some foods – usually deep-frozen types – are rendered sterile (free of disease-causing micro-organisms) by being exposed to a dose of gamma rays.

Genes The carriers of hereditary material found in the nuclei of cells.

Genus A group of closely related species; the sixth division (in descending order) in taxonomy.

Glochidium Fine hairs with tiny barbs. Also used to refer to the larvae of freshwater mussels that have barbs with which they attach to a temporary host, e.g. Bitterling (*Rhodeus* spp.), which lay their eggs in the mussels.

Glycogen A glucose-based compound which is used as a storage medium by vertebrates (backboned animals).

Gonads Sexual organs, e.g. the testes or ovaries in animals

Gonopodium Modified anal (belly) fin in male livebearing fish of the family Poeciliidae; used as a mating organ.

Gross feeders Plants, such as lilies, which obtain a significant proportion of their nutrients through their roots.

Head The height to which a pond pump can raise a column of water.

Herbivore A plant-eating animal.

Inflorescence (in flowering plants) the flower head; (in "lower" plants) the structures containing the sexual organs.

Lacustrine Of lakes.

Larvae The first stage of some fish; newly hatched invertebrates.

Metamorphosis The change in form that an animal undergoes when passing from the larval to the adult stage, e.g. the change from tadpole to froglet.

Metabolic overload Inability of an organism's metabolism to cope with excessive demands; inability to absorb sufficient oxygen for exceptionally high respiratory demand.

Micro-organism An organism whose small size makes it difficult or impossible for structural details to be clearly detected with the naked eye.

Mulch Covering consisting of straw, bark, leaves or other appropriate material used over ground or plants to combat weeds, retain moisture or offer cold-weather protection.

Mulm Organic debris, plant matter, fish waste, uneaten food, etc.

Mycobacteria Saprophytic or parasitic bacteria.

Native Plants and animals that are indigenous to an area.

Nitrate (NO_3) A compound derived from (and less toxic than) nitrite.

Nitrite (NO_2) A toxic compound derived from ammonia.

Nitrogen cycle Cyclical routes followed by nitrogen in Nature as it passes from the atmosphere, through a series of organisms and processes, until it is eventually re-released into the atmosphere.

Nitrogen fixation Binding of free nitrogen by some organisms (most notably, nitrogen-fixing bacteria) into compounds that can subsequently be absorbed by other organisms.

Omnivore An animal that eats both flesh-based and plant-based foods.

Osmosis The passage of water through a semi-permeable membrane.

Otoliths 'Stones' found in the fluid-filled ear canals of fish. Otoliths assist fish in balance control. They are also laid down in concentric rings, which can be used to determine the age of a fish.

Ovarian follicles Sacs within the ovary in which eggs (ova) are produced and sometimes retained following fertilization, e.g. as in Poeciliid livebearing fish.

pH A measure of the acid/base (acid/alkaline) properties of a solution, ranging from pH0 (extreme acidity) to pH14 (extreme alkalinity). A value of pH7 represents neutrality. The scale is logarithmic, i.e. each unit represents a 10-fold change.

Parasite An organism (plant or animal) that derives its nourishment for, at least, part of its life, from another living organism. The association usually causes some harm to the host.

Parasitic load The total number of parasites that a host carries at any particular time.

Pathogen An agent which causes disease.

Pharyngeal teeth Grinding teeth located in the throat area of many fish.

Pheromone A chemical secreted by one organism and detected by another and interpreted as a signal, e.g. the 'alarm' compounds released by fish during an attack by a predator. Unlike pheromones, hormones are maintained internally within an organism, being produced by one tissue and eliciting a response in another.

Photosynthesis A complex series of reactions in which green plants and some bacteria can convert carbon dioxide, hydrogen and oxygen into carbohydrate (food) under the influence of sunlight. Oxygen is generated as a by-product (by green plants, but not by all photosynthetic bacteria), along with water.

Phytoplankton Tiny plants which form a floating "soup" in freshwater and marine habitats worldwide.

Piscivore Fish eater.

Poikilothermic Animals whose body temperature varies with that of the surrounding medium.

Prophylaxis Treatment or procedures aimed at preventing an outbreak of disease.

Q10 A measure of how the rate of a reaction is affected by a 10°C change in temperature.

Quarantine Strictly speaking, a 40-day period of isolation. In practice, the term is generally used for any period during which, e.g. a new fish, is kept away from an established or resident specimen or population, to minimize the risk of introducing disease. Short periods are probably better referred to as "acclimatization", during which the newcomer is allowed to become used to the ambient conditions.

Respiration Complex process during which nutrients are oxidized and energy is released for metabolic purposes. In aerobic (oxygen-requiring) respiration, carbon dioxide and water are released as waste products. Some microorganisms, including de-nitrifying bacteria, can oxidize compounds, e.g. nitrates, anaerobically (in the absence of oxygen), resulting in the release of nitrogen into the atmospher.e.

Retina A "screen" at the back of the eye, onto which light rays are focussed and from which the impulses are directed to the brain, resulting in sight.

Rheophilic Living in running waters.

Rhizome A creeping, specialized stem which is produced by some plants. Rhizomes are used as storage organs and, through branching, as vegetative reproductive organs.

Riverine Of rivers.

Satellite male A male which mimics a female of the species and mates with a female while she is actually mating with a dominant male. Such males, which are about halfway in size between dominant males and sneaker males (see below) are usually able to spawn without being identified as a potential rival by the dominant male. Satellite males are quite common among some fish species e.g. sunfishes.

Sneaker male A male – usually much smaller than a dominant territorial male – which spends much of his time hidden, but keeping a close watch on a territorial male's spawning site. When a female enters the spawning area and actually begins to mate with the resident male, the sneaker "sneaks" in between the pair, releases his sperm and "sneaks" out again. As with satellites, sneaker males are quite common among the sunfishes.

Spadix Rod-like inflorescence produced by some plants, most notably the arum lilies and their relatives.

Spathe A petal-like, often colourful, bract (modified leaf) which envelops the spadix.

Species The basic unit of biological classification – any taxonomic group that a genus is broken down into. The members can interbreed.

Spermatozeugmata "Packet" of sperm produced by some male fish, e.g. Poeciliid livebearers, with which females are inseminated during copulation. Poeciliid females can store spermatozeugmata and use them to fertilize a sequence of egg batches as they ripen.

Sporangium Spore-producing organ in fungi, algae, mosses and ferns.

Stolon A creeping stem or "runner" which is capable of producing roots and stems that result in new plants.

Stratification A term usually employed in a pond keeping context to refer to the layering effect produced by rising temperature in still waters.

Turion A (usually sinking) winter bud which breaks off from the parent plant during the autumn and from which a new plant can arise the following season.

Zeolite Alumino-silicate compounds which absorb some pollutants and toxins like ammonia. Zeolite is particularly useful for coping with the ammonia surge of new ponds, which may prove overwhelming for a newly commissioned filter.

Zoonosis A disease that can be transmitted from animals to humans. The most notable one connected with pond keeping is Fish TB, which is treatable, but can prove slow to clear.

Further Reading

Allgayer, R. and J. Teton *The Complete Book of Aquarium Plants* (Ward Lock Ltd., London, UK, 1987)

Allison, J. *The Interpet Encyclopedia of Water Gardening* (Salamander Books Ltd., London, UK, 1991)

Amlacher, E. *A Textbook of Fish Diseases* (TFH Publications, Neptune City, NJ, USA, 1970)

Axelrod, H R., A. S. Benoist and D. Kelsey-Wood *The Atlas of Garden Ponds* (TFH Publications, Neptune City, NJ, USA, 1997)

Baines, C. *How to Make a Wildlife Garden* (Elm Tree Books/Hamish Hamilton Ltd., 1985)

Brewster, B., N. Chapple et al *The Interpet Practical Encyclopedia of Koi* (Salamander Books Ltd., London, UK, 1989)

Davies, R. and V. Davies (1997) *The Reptile and Amphibian Problem Solver* (Tetra Press, Blacksburg, VA, USA, 1997)

Dawes, J. *John Dawes's Book of Water Gardens* (TFH Publications, Neptune City, NJ, USA, 1989)

Dawes, J. *Tropical Aquarium Fish* (New Holland Publications Ltd., 1996)

Dawes, J. *Livebearing Fishes – A Guide to their Aquarium Care, Biology and Classification* (Blandford, London, UK 1991, 1995)

James, B. *A Fishkeeper's Guide to Koi* (Salamander Books Ltd., London, UK, 1986)

Mills, D. *A Fishkeeper's Guide to Coldwater Fishes* (Salamander Books Ltd., London, UK, 1984)

Mühlberg, H. *The Complete Guide to Water Plants* (EP Publishing Ltd., 1982)

Pool, D. *Hobbyist Guide to Successful Koi Keeping* (Tetra Press, Blacksburg, VA, USA, 1991)

Robinson, P. *The Royal Horticultural Society – Water Gardening* (Dorling Kindersley Ltd., London, UK, 1997)

Sandford, G. *The Tropical Freshwater Aquarium Problem Solver* (Tetra Press, Blacksburg, VA, USA, 1998)

Sandford, G. *An Illustrated Encyclopedia of Aquarium Fish* (Quintet Publishing Ltd., London, UK, 1995)

Stadelmann, P. *Water Gardens* (Barron's Educational Series, Hauppauge, NY, USA, 1992)

Stein, S. *Water Gardening* (Grange Books, London, UK, 1994)

Swindells, P. *Waterlilies* (Croom Helm Ltd., 1983)

Swindells, P. *The Master Book of Water Gardens* (Salamander Books Ltd., London, UK, 1997)

Swindells, P. *Water Gardening for Everyone* (Burall Floraprint Ltd., Wisbech, UK, 1983)

Acknowledgments

Abbreviations

AOL	Andromeda Oxford Limited
DB	Dave Bevan
GPL	Garden Picture Library
HS	Harry Smith Collection
JB	Jane Burton
JG	John Glover Photography
Lotus	By kind permission of Lotus Water Garden Products Limited
OASE	Reproduced with kind permission of Roger Cox at OASE (UK) Ltd
PM	Photomax, Max Gibbs

1 HS; 2 AOL; 3tl Hozelock Cyprio; 3tr Lotus; 3b PM; 7 Andrew Lawson; 8–9 Steven Wooster/GPL; 9 JG; 10 Vaughan Fleming/GPL; 11 JB; 12–13 Salamander Picture Library; 15cr JB; 15br Practical Fishkeeping; 17 HS; 21t JG; 21b HS; 22 JG; 23 Lotus; 24 JG; 24–25 HS; 27 Brigitte Thomas/GPL; 28–29 HS; 31 Gil Hanly/GPL; 32–33 JG; 34l, 34r, 35b Lotus; 35t JG; 38 Jerry Harpur/Designed by Ernie Taylor; 39 AOL; 41 Martin Anderson/AOL; 44t, 44b, 45l Lotus; 45fl, 45r, 45fr, 46t, 46c Hozelock Cyprio; 47 Interpet Ltd; 48, 49 AOL; 50 (clockwise from tl) OASE; Lotus; Lotus; OASE; Hozelock Cyprio; Lotus; 51 Jerry Harpur/Designed by Jean Melville-Clark; 52 Interpet Ltd; 53t Trident Water Garden Products Ltd; 53c OASE; 53b Hozelock Cyprio; 55 Eric Crichton/Bruce Coleman Limited; 56 John Dawes; 57 Martin Anderson/AOL; 58 PM; 60, 61 Martin Anderson/AOL; 63 JB; 64–65 PM; 66–67 AOL; 69 PM; 70–71, 71 Kim Taylor; 72, 73, 75t, 75b PM; 76t Michael Edwards; 76b PM; 77 DB; 78tl Dr. Chris Andrews/South Carolina Aquarium; 78tr, 78b, 82–83 PM; 83 JB; 85l, 85r, 86–87 PM; 88–89 Lythgoe/Planet Earth Pictures; 90c DB; 90b PM; 92 Michael Edwards; 93 PM; 94–95 JB; 96, 97, 98, 99, 101t, 101c, 101b PM; 102 JB; 103 DB; 104–105 PM; 106 DB; 107 PM; 109 MP. and C.

Piednoir/Aquapress; 110, 111 JB; 114–115, 115 PM; 117 JG; 118l JB; 118r HS; 120 Heather Angel/Biofotos; 121 DB; 126 HS; 127 Quadrillion Publishing; 128 DB; 129 Heather Angel/Biofotos; 131 Steven Wooster/GPL; 132 HS; 133 Sunniva Harte; 134 HS; 135t, 135b AOL; 136 HS; 137 Jerry Harpur/Designed by Chris Grey-Wilson; 138, 139 HS; 140t Steven Wooster/GPL; 140b HS; 141 Howard Rice/GPL; 142 Didier Willery/GPL; 143 HS; 144 Didier Willery/GPL; 145 AOL; 146 Ron Sutherland/GPL; 147 Rex Butcher/GPL; 148 Gary Rogers/GPL; 149 Brigitte Thomas/GPL; 150–151 Jerry Harpur/Odiham Nursery; 152, 153t Howard Rice/GPL; 153b Rex Butcher/GPL; 154, 155 HS; 156 Jerry Harpur/Designed by Craig and Donna Rothman; 157 Sunniva Harte/GPL; 158 DB; 159 HS; 160 Jerry Harpur/Great Dixter; 161, 163 JB; 164–165 Harry Girier/FTFFA/John Dawes; 165 JB; 166–167 HS; 168 David M. Dennis; 169, 171 JB; 172 Sunniva Harte; 173, 174 JB; 175 Howard Rice/GPL; 176 John Dawes; 176–177 Gary Rogers/GPL; 179 Heather Angel/Biofotos; 181 JB; 182–183 Nigel Francis/GPL; 183 Diplex Limited; 185 Michael Edwards; 186 HS; 187 AOL; 188 JG; 189 Lamontagne/GPL; 190 HS; 191l Dr. J.G. Bateman; 191r JG; 193 Lamontagne/GPL; 194–195 Jerry Harpur/Designed by Jean Melville-Clark; 196 Chris Munday; 197 JB; 198 Didier Willery/GPL; 198–199 Michael Howes/GPL

Artwork by Julian Baker, Richard Lewington, Michael Loates, Denys Ovenden and Graham Rosewarne.

The authors and publishers would like to thank the following people for their help with this project: Anthony Atkinson & Alex Firkin of Appleby Constructions; John & Sheila Coyle; Fawcett's Liners, Longton, Lancs. (supply of coloured liners on p. 39); Millets Farm Garden Centre, Frilford, nr Abingdon, Oxfordshire, and Ray & June Tolley (assistance with fish release sequence on pp. 60–61); and World of Water Aquatic Centres (loan of equipment and materials).

Index

Page numbers in *italics* refer to picture captions. Numbers in **bold** refer to a major treatment of a subject.

Index